Biology of Marine Mammals

Biology of Marine Mammals

Edited by
Hyman Dunn

Larsen & Keller
www.larsen-keller.com

Biology of Marine Mammals
Edited by Hyman Dunn
ISBN: 978-1-63549-174-6 (Hardback)

© 2017 Larsen & Keller

☰ Larsen & Keller

Published by Larsen and Keller Education,
5 Penn Plaza,
19th Floor,
New York, NY 10001, USA

Cataloging-in-Publication Data

Biology of marine mammals / edited by Hyman Dunn.
      p. cm.
Includes bibliographical references and index.
ISBN 978-1-63549-174-6
      1. Marine mammals. 2. Marine biology. 3. Aquatic mammals.
I. Dunn, Hyman.
QL713.2 .B56 2017
599.5--dc23

The publisher's policy is to use permanent paper from mills that operate a sustainable forestry policy. Furthermore, the publisher ensures that the text paper and cover boards used have met acceptable environmental accreditation standards.

Printed and bound in the United States of America.

For more information regarding Larsen and Keller Education and its products, please visit the publisher's website www.larsen-keller.com

Table of Contents

Preface

Marine mammal biology studies mammals such as whales, seals, polar bears, etc. that rely on the aquatic ecosystem for their existence. These animals are grouped together as they have a paraphyletic relation or share a common ancestor. This book is a valuable compilation of topics, ranging from the basic to the most complex theories and principles in the field of marine mammal biology. Also included in this text is a detailed explanation of the various concepts and insights of aquatic ecosystem. The contents of this text are of utmost significance and are bound to provide incredible insights to the readers.

To facilitate a deeper understanding of the contents of this book a short introduction of every chapter is written below:

Chapter 1- Marine mammals are mammals that are found in the ocean. Marine mammals include animals such as seals, sea otters and polar bears. The chapter on marine mammals offers an insightful focus, keeping in mind the subject matter.

Chapter 2- Aquatic mammals are animals that live partly or entirely in water bodies. They include animals such as the European otter and the Amazon river dolphin. Blubber, cetacean bycatch, aquatic locomotion, marine mammal observer and marine mammals and sonar are some aspects of aquatic mammals. This section is an overview of the subject matter incorporating all the major aspects of aquatic mammal.

Chapter 3- The various species of marine mammals are sirenia, manatee, pinniped, sea otter and polar bears. Manatees are large mammals that are fully aquatic and are also sometimes known as sea cows whereas pinnipeds are sometimes known as seals and are semiaquatic marine mammals. The aspects elucidates in this chapter are of vital importance, and provide a better understanding of marine mammals.

Chapter 4- A whale is a diverse group of aquatic mammals; whales encompass eight extant families. Baleen whales and marine otters have also been explained in the section. The chapter on whales and marine otters offers an insightful focus, keeping in mind the subject matter.

Chapter 5- The sea mink has become an extinct marine mammal; they had a distinctive odor and had fur that was known to be rough. The other extinct marine mammals explained in the chapter are Steller's sea cow, Japanese sea lion and Caribbean monk seal. This section helps the readers in understanding the causes of disappearance of these marine mammals.

I owe the completion of this book to the never-ending support of my family, who supported me throughout the project.

Editor

Introduction to Marine Mammals

Marine mammals are mammals that are found in the ocean. Marine mammals include animals such as seals, sea otters and polar bears. The chapter on marine mammals offers an insightful focus, keeping in mind the subject matter.

Marine mammals are aquatic mammals that rely on the ocean and other marine ecosystems for their existence. They include animals such as seals, whales, manatees, sea otters and polar bears. They do not represent a distinct taxon or systematic grouping, but rather have a paraphyletic relation. They are also unified by their reliance on the marine environment for feeding.

A humpback whale (*Megaptera novaeangliae*), a member of infraorder Cetacea of the order Cetartiodactyla

Marine mammal adaptation to an aquatic lifestyle vary considerably between species. Both cetaceans and sirenians are fully aquatic and therefore are obligate water dwellers. Seals and sea-lions are semiaquatic; they spend the majority of their time in the water, but need to return to land for important activities such as mating, breeding and molting. In contrast, both otters and the polar bear are much less adapted to aquatic living. Their diet varies considerably as well; some may eat zooplankton, others may eat fish, squid, shellfish, sea-grass and a few may eat other mammals. While the number of marine mammals is small compared to those found on land, their roles in various ecosystems are large, especially concerning the maintenance of marine ecosystems, through processes including the regulation of prey populations. This role in maintaining ecosystems makes them of particular concern as 23% of marine mammal species are currently threatened.

Marine mammals were first hunted by aboriginal peoples for food and other resources. Many were also the target for commercial industry, leading to a sharp decline in all populations of exploited species, such as whales and seals. Commercial hunting lead to the extinction of †Steller's sea cow and the †Caribbean monk seal. After commercial hunting ended, some species, such as the gray whale and northern elephant seal, have rebounded in numbers; conversely, other species, such as the North Atlantic right whale, are critically endangered. Other than hunting, marine mammals can be killed as bycatch from fisheries, where they become entangled in fixed netting and drown or starve. Increased ocean traffic causes collisions between fast ocean vessels and large marine mammals. Habitat degradation also threatens marine mammals and their ability to find and catch food. Noise pollution, for example, may adversely affect echolocating mammals, and the ongoing effects of global warming degrade arctic environments.

A leopard seal (Hydrurga leptonyx), a member of the clade Pinnipedia of the order Carnivora

Taxonomy

Marine mammals form a diverse group of 129 species that rely on the ocean for their existence. They do not represent a distinct taxon or systematic grouping, but instead have a paraphyletic relationship. They are also unified by their reliance on the marine environment for feeding. The level of dependence on the marine environment for existence varies considerably with species. For example, dolphins and whales are completely dependent on the marine environment for all stages of their life, seals feed in the ocean but breed on land, and polar bears must feed on land. Twenty three percent of marine mammal species are threatened.

Marine mammals vary greatly in size and shape		

A polar bear (*Ursus maritimus*), a member of family Ursidae.	A sea otter (*Enhydra lutris*), a member of family Mustelidae.	California sea lions (*Zalophus californianus*), members of the family Otariidae.

Classification of Extant Species

- Order Cetartiodactyla
  - Suborder Whippomorpha
    - Family Balaenidae (right and bowhead whales), two genera and four species
    - Family Cetotheriidae (pygmy right whale), one species
    - Family Balaenopteridae (rorquals), two genera and eight species
    - Family Eschrichtiidae (gray whale), one species
    - Family Physeteridae (sperm whale), one species
    - Family Kogiidae (pygmy and dwarf sperm whales), one genus and two species
    - Family Monodontidae (narwhal and beluga), two genera and two species
    - Family Ziphiidae (beaked whales), six genera and 21 species
    - Family Delphinidae (oceanic dolphins), 17 genera and 38 species
    - Family Phocoenidae (porpoises), two genera and seven species
- Order Sirenia (sea cows)
  - Suborder Cynodontia
    - Family Trichechidae (manatees), one species
    - Family Dugongidae (dugongs), one species
- Order Carnivora (carnivores):
  - Suborder Caniformia
    - Family Mustelidae, two species
    - Family Ursidae (bears), one species
  - Suborder Pinnipedia (sealions, walruses, seals)
    - Family Otariidae (eared seals), seven genera and 15 species
    - Family Odobenidae (walrus), one species
    - Family Phocidae (earless seals), 14 genera and 18 species

Evolution

Mammals have returned to the ocean in many separate evolutionary lineages, namely: Cetacea, Sirenia, †Desmostylia, Pinnipedia, †*Kolponomos* (marine bear), †*Thalassocnus* (aquatic sloth), *Ursus maritimus* (polar bear), and *Enhydra lutris* (sea otter); the eutriconodonts †*Ichthyoconodon* and †*Dyskritodon* might have also been marine in habits. Five of these lineages are extinct (†Desmostylia; †*Kolponomos*; †*Thalassocnus*, †*Dyskritodon*, †*Ichthyoconodon*). Despite the diversity in morphology seen between groups, improving foraging efficiency has been the main driver in the evolution in these lineages. Today, fully aquatic marine mammals belong to one of two orders: Cetartiodactyla or Sirenia. The Cetartiodactyla lineage became aquatic around 50 million years ago (mya), then Sirenia 40 mya, and Pinnipedia around 20 to 25 mya, then sea otters two mya, and most recently the polar bear around 130,000 to 110,000 years ago.

A skeleton of †*Thalassocnus* (5–3 mya) from the Muséum national d'histoire naturelle in its presumed swimming pose

Based on molecular and morphological research, the cetaceans genetically and morphologically fall firmly within the Artiodactyla (even-toed ungulates). The term Cetartiodactyla reflects the idea that whales evolved within the ungulates. The term was coined by merging the name for the two orders, Cetacea and Artiodactyla, into a single word. Under this definition, the closest living land relative of the whales and dolphins is thought to be the hippopotamuses. Use of the order Cetartiodactyla, instead of Cetacea with parvorders Odontoceti and Mysticeti, is favored by most evolutionary mammalogists working with molecular data and is supported the IUCN Cetacean Specialist Group and by Taxonomy Committee of the Society for Marine Mammalogy, the largest international association of marine mammal scientists in the world. Some others, including many marine mammalogists and paleontologists, favor retention of order Cetacea with the two suborders in the interest of taxonomic stability.

Pinnipeds split from other caniforms 50 mya during the Eocene. Their evolutionary link to terrestrial mammals was unknown until the 2007 discovery of †*Puijila darwini* in early Miocene deposits in Nunavut, Canada. Like a modern otter, †*Puijila* had a long tail, short limbs and webbed feet instead of flippers. The lineages of Otariidae (eared seals) and Odobenidae (walrus) split almost 28 mya. Phocids (earless seals) are known to have existed for at least 15 mya, and molecular evidence supports a divergence of the Monachinae (monk seals) and Phocinae lineages 22 mya.

Fossil evidence indicates the sea otter (*Enhydra*) lineage became isolated in the North Pacific approximately two mya, giving rise to the now-extinct †*Enhydra macrodonta* and the modern sea otter, *Enhydra lutris*. The sea otter evolved initially in northern Hokkaidō and Russia, and then spread east to the Aleutian Islands, mainland Alaska, and down the North American coast. In comparison to cetaceans, sirenians, and pinnipeds, which entered the water approximately 50, 40, and 20 mya, respectively, the sea otter is a relative newcomer to marine life. In some respects though, the sea otter is more fully adapted to water than pinnipeds, which must haul out on land or ice to give birth.

Illustration of †*Prorastomus*, the most known primitive sirenian (40 mya)

The first appearance of sirenians in the fossil record was during the early Eocene, and by the late Eocene, sirenians had significantly diversified. Inhabitants of rivers, estuaries, and nearshore marine waters, they were able to spread rapidly. The most primitive sirenian, †*Prorastomus*, was found in Jamaica, unlike other marine mammals which originated from the Old World (such as cetaceans). The first known quadrupedal sirenian was †*Pezosiren* from the early Eocene. The earliest known sea cows, of the families †Prorastomidae and †Protosirenidae, were both confined to the Eocene, and were pig-sized, four-legged, amphibious creatures. The first members of Dugongidae appeared by the end of the Eocene. At this point, sea cows were fully aquatic.

Polar bears are thought to have diverged from a population of brown bears, *Ursus arctos*, that became isolated during a period of glaciation in the Pleistocene or from the eastern part of Siberia, (from Kamchatka and the Kolym Peninsula). The oldest known polar bear fossil is a 130,000 to 110,000-year-old jaw bone, found on Prince Charles Foreland in 2004. The mitochondrial DNA (mtDNA) of the polar bear diverged from the brown bear roughly 150,000 years ago. Further, some clades of brown bear, as assessed by their mtDNA, are more closely related to polar bears than to other brown bears, meaning that the polar bear might not be considered a species under some species concepts.

Distribution and Habitat

Marine mammals are widely distributed throughout the globe, but their distribution is patchy and coincides with the productivity of the oceans. Species richness peaks at around 40° latitude, both north and south. This corresponds to the highest levels of primary production around North and South America, Africa, Asia and Australia. Total species range is highly variable for marine mammal species. On average most marine mammals have ranges which are equivalent or smaller than one-fifth of the Indian Ocean. The variation observed in range size is a result of the different ecological requirements of each species and their ability to cope with a broad range of environmental conditions. The high degree of overlap between marine mammal species richness and areas of human impact on the environment is of concern.

Most marine mammals, such as seals and sea otters, inhabit the coast. Seals, however, also use a number of terrestrial habitats, both continental and island. In temperate and tropical areas, they haul-out on to sandy and pebble beaches, rocky shores, shoals, mud flats, tide pools and in sea caves. Some species also rest on man-made structures, like piers, jetties, buoys and oil platforms. Seals may move further inland and rest in sand dunes or vegetation, and may even climb cliffs. Most cetaceans live in the open ocean, and species like the sperm whale may dive to depths of −1,000 to −2,500 feet (−300 to −760 m) in search of food. Sirenians live in shallow coastal waters, usually living 30 feet (9.1 m) below sea level. However, they have been known to dive to −120 feet (−37 m) to forage deep-water seagrasses. Sea otters live in protected areas, such as rocky shores, kelp forests, and barrier reefs, although they may reside among drift ice or in sandy, muddy, or silty areas.

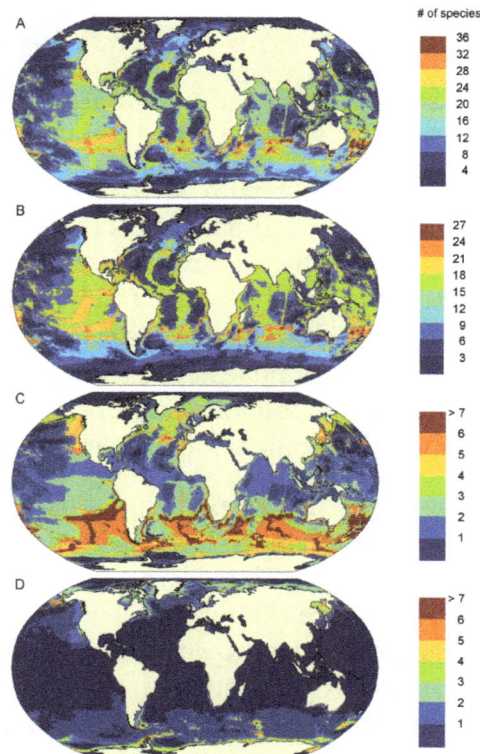

Predicted patterns of marine mammal species richness. A) All species (n=115), B) Odontocetes (n=69), C) Mysticetes (n=14), D) Pinnipeds (n=32). Colors indicate the number of species predicted to occur in each 0.5°x0.5° grid cell from a relative environmental suitability (RES) model, using environmental data from 1990–1999, and assuming a presence threshold of RES>0.6.

Many marine mammals seasonally migrate. Annual ice contains areas of water that appear and disappear throughout the year as the weather changes, and seals migrate in response to these changes. In turn, polar bears must follow their prey. In Hudson Bay, James Bay, and some other areas, the ice melts completely each summer (an event often referred to as "ice-floe breakup"), forcing polar bears to go onto land and wait through the months until the next freeze-up. In the Chukchi and Beaufort seas, polar bears retreat each summer to the ice further north that remains frozen year-round. Seals may also migrate to other environmental changes, such as El Niño, and traveling seals may use various features of their environment to reach their destination including geomagnetic fields, water and wind currents, the position of the sun and moon and the taste and temperature of the water. Baleen whales famously migrate very long distances into tropical waters to give birth and raise young, possibly to prevent predation by killer whales. The gray whale has the longest recorded migration of any mammal, with one traveling 14,000 miles (23,000 km) from the Sea of Okhotsk to the Baja Peninsula. During the winter, manatees living at the northern end of their range migrate to warmer waters.

Adaptations

Anatomy of a Dolphin
(Delphinidae)

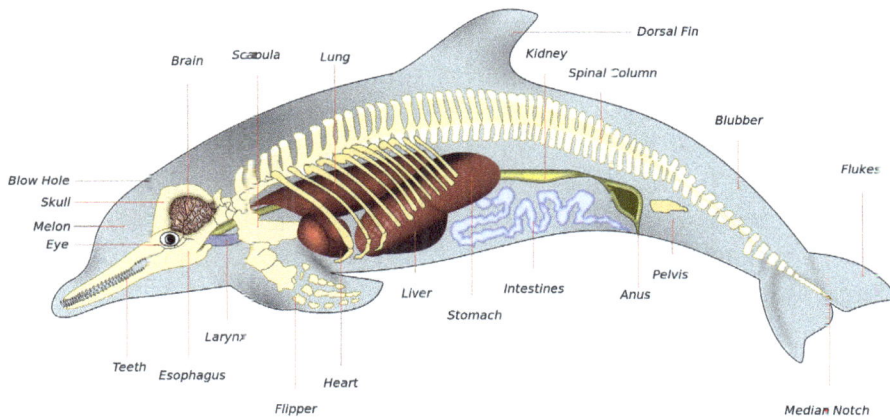

The anatomy of a dolphin showing its skeleton, major organs, and body shape

Marine mammals have a number of physiological and anatomical features to overcome the unique challenges associated with aquatic living. Some of these features are very species specific. Marine mammals have developed a number of features for efficient locomotion such as torpedo shaped bodies to reduce drag; modified limbs for propulsion and steering; tail flukes and dorsal fins for propulsion and balance. Marine mammals are adept at thermoregulation using dense fur or blubber, circulatory adjustments (counter-current heat exchangers); torpedo shaped bodies, reduced appendages, and large size to prevent heat loss.

Marine mammals are able to dive for long periods of time. Both pinnipeds and cetaceans have large and complex blood vessel systems which serve to store oxygen to support deep diving. Other important reservoirs include muscles, blood, and the spleen which all have the capacity to hold a high concentration of oxygen. They are also capable of bradycardia (reduced heart rate), and vasoconstriction (shunting most of the oxygen to vital organs such as the brain and heart) to allow extended diving times and cope with oxygen deprivation. If oxygen is depleted, marine mammals can access substantial reservoirs of glycogen that support anaerobic glycolysis of the cells involved during conditions of systemic hypoxia associated with prolonged submersion.

Sound travels differently through water, and therefore marine mammals have developed adaptations to ensure effective communication, prey capture, and predator detection. The most notable adaptation is the development of echolocation in whales and dolphins. Toothed whales emit a focused beam of high-frequency clicks in the direction that their head is pointing. Sounds are generated by passing air from the bony nares through the phonic lips.[p. 112] These sounds are reflected by the dense concave bone of the cranium and an air sac at its base. The focused beam is modulated by a large fatty organ known as the 'melon'. This acts like an acoustic lens because it is composed of lipids of differing densities.

Marine mammals have evolved a wide variety of features for feeding, which are mainly seen in their dentition. For example, the cheek teeth of pinnipeds and odontocetes are specifically adapted to capture fish and squid. In contrast, baleen whales have evolved baleen plates to filter feed plankton and small fish from the water.

Polar bears, otters, and fur seals have fur, one of the defining mammalian features, that is long, oily, and waterproof in order to trap air to provide insulation. In contrast, other marine mammals – such as whales, dolphins, porpoises, manatees, dugongs, and walruses – have lost long fur in favor of a thick, dense epidermis and a thickened fat layer (blubber) in response to hydrodynamic requirements. Wading and bottom-feeding animals (such as manatee) need to be heavier than water in order to keep contact with the floor or to stay submerged. Surface-living animals (such as sea otters) need the opposite, and free-swimming animals living in open waters (such as dolphins) need to be neutrally buoyant in order to be able to swim up and down the water column. Typically, thick and dense bone is found in bottom feeders and low bone density is associated with mammals living in deep water. Some marine mammals, such as polar bears and otters, have retained four weight-bearing limbs and can walk on land like fully terrestrial animals.

Ecology

Diet

Killer whale hunting a Weddel seal

All cetaceans are carnivorous and predatory. Toothed whales mostly feed on fish and cephalopods, followed by crustaceans and bivalves. Some may forage with other kinds of animals, such as other species of whales or certain species of pinnipeds. One common feeding method is herding, where a pod squeezes a school of fish into a small volume, known as a bait ball. Individual members then take turns plowing through the ball, feeding on the stunned fish. Coralling is a method where dolphins chase fish into shallow water to catch them more easily. Killer whales and bottlenose dolphins have also been known to drive their prey onto a beach to feed on it. Other whales with a blunt snout and reduced dentition rely on suction feeding. Though carnivorous, they house gut flora similar to that of terrestrial herbivores, probably a remnant of their herbivorous ancestry.

Baleen whales use their baleen plates to sieve plankton, among others, out of the water; there are two types of methods: lunge-feeding and gulp-feeding. Lunge-feeders expand the volume of their jaw to a volume bigger than the original volume of the whale itself by inflating their mouth. This causes grooves on their throat to expand, increasing the amount of water the mouth can store. They ram a baitball at high speeds in order to feed, but this is only energy-effective when used against a large baitball. Gulp-feeders swim with an open mouth, filling it with water and prey. Prey must occur in sufficient numbers to trigger the whale's interest, be within a certain size range so that the baleen plates can filter it, and be slow enough so that it cannot escape.

Sea otters have dexterous hands which they use to smash sea urchins off of rocks.

Otters are the only marine animals that are capable of lifting and turning over rocks, which they often do with their front paws when searching for prey. The sea otter may pluck snails and other organisms from kelp and dig deep into underwater mud for clams. It is the only marine mammal that catches fish with its forepaws rather than with its teeth. Under each foreleg, sea otters have a loose pouch of skin that extends across the chest which they use to store collected food to bring to the surface. This pouch also holds a rock that is used to break open shellfish and clams, an example of tool use. The sea otters eat while floating on their backs, using their forepaws to tear food apart and bring to their mouths. Marine otters mainly feed on crustaceans and fish.

Pinnipeds mostly feed on fish and cephalopods, followed by crustaceans and bivalves, and then zooplankton and warm-blooded prey (like sea birds). Most species are generalist feeders, but a few are specialists. They typically hunt non-schooling fish, slow-moving or immobile invertebrates or endothermic prey when in groups. Solitary foraging species usually exploit coastal waters, bays and rivers. When large schools of fish or squid are available, pinnipeds hunt cooperatively in large groups, locating and herding their prey. Some species, such as California and South American sea lions, may forage with cetaceans and sea birds.

The polar bear is the most carnivorous species of bear, and its diet primarily consists of ringed (*Pusa hispida*) and bearded (*Erignathus barbatus*) seals. Polar bears hunt primarily at the inter-face between ice, water, and air; they only rarely catch seals on land or in open water. The polar bear's most common hunting method is still-hunting: The bear locates a seal breathing hole using its sense of smell, and crouches nearby for a seal to appear. When the seal exhales, the bear smells its breath, reaches into the hole with a forepaw, and drags it out onto the ice. The polar bear also hunts by stalking seals resting on the ice. Upon spotting a seal, it walks to within 100 yards (90 m), and then crouches. If the seal does not notice, the bear creeps to within 30 to 40 feet (9 to 10 m) of the seal and then suddenly rushes to attack. A third hunting method is to raid the birth lairs that female seals create in the snow. They may also feed on fish.

A dugong feeding on the sea-floor

Sirenians are referred to as "sea cows" because their diet consists mainly of sea-grass. When eating, they ingest the whole plant, including the roots, although when this is impossible they feed on just the leaves. A wide variety of seagrass has been found in dugong stomach contents, and evidence exists they will eat algae when seagrass is scarce. West Indian manatees eat up to 60 different species of plants, as well as fish and small invertebrates to a lesser extent.

Keystone Species

Sea otters are a classic example of a keystone species; their presence affects the ecosystem more profoundly than their size and numbers would suggest. They keep the population of certain benthic (sea floor) herbivores, particularly sea urchins, in check. Sea urchins graze on the lower stems of kelp, causing the kelp to drift away and die. Loss of the habitat and nutrients provided by kelp forests leads to profound cascade effects on the marine ecosystem. North Pacific areas that do not have sea otters often turn into urchin barrens, with abundant sea urchins and no kelp forest. Reintroduction of sea otters to British Columbia has led to a dramatic improvement in the health of coastal ecosystems, and similar changes have been observed as sea otter populations recovered in the Aleutian and Commander Islands and the Big Sur coast of California However, some kelp forest ecosystems in California have also thrived without sea otters, with sea urchin populations apparently controlled by other factors. The role of sea otters in maintaining kelp forests has been observed to be more important in areas of open coast than in more protected bays and estuaries.

Antarctic fur seal pups (left) vs. Arctic harp seal pup (right)

An apex predator affects prey population dynamics and defense tactics (such as camouflage). The polar bear is the apex predator within its range. Several animal species, particularly Arctic foxes (*Vulpes lagopus*) and glaucous gulls (*Larus hyperboreus*), routinely scavenge polar bear kills. The relationship between ringed seals and polar bears is so close that the abundance of ringed seals in some areas appears to regulate the density of polar bears, while polar bear predation in turn regulates density and reproductive success of ringed seals. The evolutionary pressure of polar bear predation on seals probably accounts for some significant differences between Arctic and Antarctic seals. Compared to the Antarctic, where there is no major surface predator, Arctic seals use more breathing holes per individual, appear more restless when hauled out on the ice, and rarely defecate on the ice. The fur of Arctic pups is white, presumably to provide camouflage from predators, whereas Antarctic pups all have dark fur.

Killer whales are apex predators throughout their global distribution, and can have a profound effect on the behavior and population of prey species. Their diet is very broad and they can feed on many vertebrates in the ocean including salmon, rays, sharks (even white sharks), large baleen whales, and nearly 20 species of pinniped. The predation of whale calves may be responsible for annual whale migrations to calving grounds in more tropical waters, where the population of killer whales is much lower than in polar waters. Prior to whaling, it is thought that great whales were a major food source; however, after their sharp decline, killer whales have since expanded their diet, leading to the decline of smaller marine mammals. A decline in Aleutian Islands sea otter populations in the 1990s was controversially attributed by some scientists to killer whale predation, although with no direct evidence. The decline of sea otters followed a decline in harbor seal and Steller sea lion populations, the killer whale's preferred prey, which in turn may be substitutes for their original prey, now reduced by industrial whaling.

Whale Pump

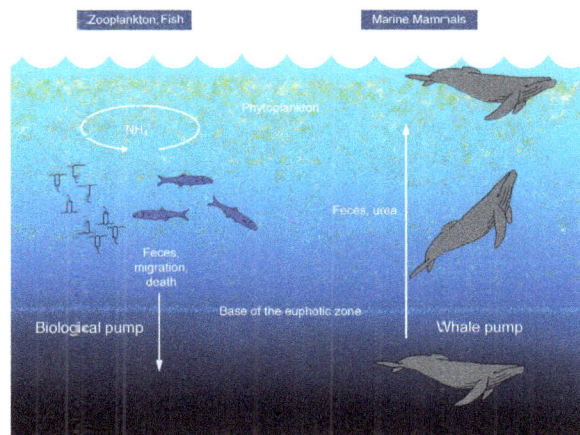

"Whale pump" – the role played by whales in recycling ocean nutrients

A 2010 study considered whales to be a positive influence to the productivity of ocean fisheries, in what has been termed a "whale pump". Whales carry nutrients such as nitrogen from the depths back to the surface. This functions as an upward biological pump, reversing an earlier presumption that whales accelerate the loss of nutrients to the bottom. This nitrogen input in the Gulf of Maine is more than the input of all rivers combined emptying into the gulf, some 25,000 short tons (23,000 t) each year. Whales defecate at the ocean's surface; their excrement is important for fisheries because it is rich in iron and nitrogen. The whale feces are liquid and instead of sinking, they stay at the surface where phytoplankton feed off it.

Upon death, whale carcasses fall to the deep ocean and provide a substantial habitat for marine life. Evidence of whale falls in present-day and fossil records shows that deep sea whale falls support a rich assemblage of creatures, with a global diversity of 407 species, comparable to other neritic biodiversity hotspots, such as cold seeps and hydrothermal vents. Deterioration of whale carcasses happens though a series of three stages. Initially, moving organisms, such as sharks and hagfish, scavenge soft tissue at a rapid rate over a period of months to as long as two years. This is followed by the colonization of bones and surrounding sediments (which contain organic matter) by enrichment opportunists, such as crustaceans and polychaetes, throughout a period of years. Finally, sulfophilic bacteria reduce the bones releasing hydrogen sulphide enabling the growth of chemoautotrophic organisms, which in turn, support other organisms such as mussels, clams, limpets, and sea snails. This stage may last for decades and supports a rich assemblage of species, averaging 185 species per site.

Interactions with Humans

Threats

Exploitation

Men killing northern fur seals on Saint Paul Island, Alaska in the 1890s

Marine mammals were hunted by coastal aboriginal humans historically for food and other resources. These subsistence hunts still occur in Canada, Greenland, Indonesia, Russia, the United States, and several nations in the Caribbean. The effects of these are only localized, as hunting efforts were on a relatively small scale. Commercial hunting took this to a much greater scale and marine mammals were heavily exploited. This led to the extinction of the †Steller's sea cow (along with subsistence hunting) and the †Caribbean monk seal. Today, populations of species that were historically hunted, such as blue whales (*Balaenoptera musculus*) and the North Pacific right whale (*Eubalaena japonica*), are much lower than their pre-whaling levels. Because whales generally have slow growth rates, are slow to reach sexual maturity, and have a low reproductive output, population recovery has been very slow.

A number of whales are still subject to direct hunting, despite the 1986 moratorium ban on whaling set under the terms of the International Whaling Commission (IWC). There are only two nations remaining which sanction commercial whaling: Norway, where several hundred common minke whales are harvested each year; and Iceland, where quotas of 150 fin whales and 100 minke whales per year are set. Japan also harvests several hundred Antarctic and North Pacific minke whales each year, ostensibly for scientific research in accordance with the moratorium. However, the illegal trade of whale and dolphin meat is a significant market in Japan and some countries.

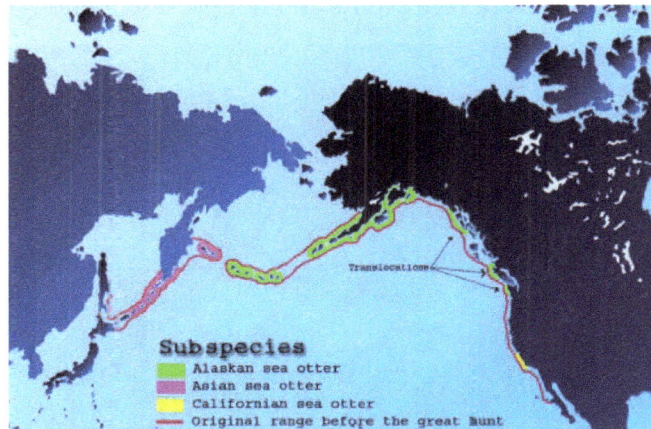
Historical and modern range of northern sea otters

The most profitable furs in the fur trade were those of sea otters, especially the northern sea otter which inhabited the coastal waters between the Columbia River to the south and Cook Inlet to the north. The fur of the Californian southern sea otter was less highly prized and thus less profitable. After the northern sea otter was hunted to local extinction, maritime fur traders shifted to California until the southern sea otter was likewise nearly extinct. The British and American maritime fur traders took their furs to the Chinese port of Guangzhou (Canton), where they worked within the established Canton System. Furs from Russian America were mostly sold to China via the Mongolian trading town of Kyakhta, which had been opened to Russian trade by the 1727 Treaty of Kyakhta.

Commercial sealing was historically just as important as the whaling industry. Exploited species included harp seals, hooded seals, Caspian seals, elephant seals, walruses and all species of fur seal. The scale of seal harvesting decreased substantially after the 1960s, after the Canadian government reduced the length of the hunting season and implemented measures to protect adult females. Several species that were commercially exploited have rebounded in numbers; for example, Antarctic fur seals may be as numerous as they were prior to harvesting. The northern elephant seal was hunted to near extinction in the late 19th century, with only a small population remaining on Guadalupe Island. It has since recolonized much of its historic range, but has a population bottleneck. Conversely, the Mediterranean monk seal was extirpated from much of its former range, which stretched from the Mediterranean to the Black Sea and northwest Africa, and only remains in the northeastern Mediterranean and some parts of northwest Africa.

Ocean Traffic and Fisheries

By-catch is the incidental capture of non-target species in fisheries. Fixed and drift gill nets cause the highest mortality levels for both cetaceans and pinnipeds, however, entanglements in long lines, mid-water trawls, and both trap and pot lines are also common. Tuna seines are particularly problematic for entanglement by dolphins. By-catch affects all cetaceans, both small and big, in all habitat types. However, smaller cetaceans and pinnipeds are most vulnerable as their size means that escape once they are entangled is highly unlikely and they frequently drown. While larger cetaceans are capable of dragging nets with them, the nets sometimes remain tightly attached to the individual and can impede the animal from feeding sometimes leading to starvation. Abandoned or lost nets and lines cause mortality through ingestion or entanglement. Marine mammals also get entangled in aquaculture nets, however, these are rare events and not prevalent enough to impact populations.

The remains of a North Atlantic right whale after it collided with a ship propeller

Vessel strikes cause death for a number of marine mammals, especially whales. In particular, fast commercial vessels such as container ships can cause major injuries or death when they collide with marine mammals. Collisions occur both with large commercial vessels and recreational boats and cause injury to whales or smaller cetaceans. The critically endangered North Atlantic right whale is particularly affected by vessel strikes. Tourism boats designed for whale and dolphin watching can also negatively impact on marine mammals by interfering with their natural behavior.

The fishery industry not only threatens marine mammals through by-catch, but also through competition for food. Large scale fisheries have led to the depletion of fish stocks that are important prey species for marine mammals. Pinnipeds have been especially affected by the direct loss of food supplies and in some cases the harvesting of fish has led to food shortages or dietary deficiencies, starvation of young, and reduced recruitment into the population. As the fish stocks have been depleted, the competition between marine mammals and fisheries has sometimes led to conflict. Large-scale culling of populations of marine mammals by commercial fishers has been initiated in a number of areas in order to protect fish stocks for human consumption.

Shellfish aquaculture takes up space so in effect creates competition for space. However, there is little direct competition for aquaculture shellfish harvest. On the other hand, marine mammals regularly take finfish from farms, which creates significant problems for marine farmers. While there are usually legal mechanisms designed to deter marine mammals, such as anti-predator nets or harassment devices, individuals are often illegally shot.

Habitat Loss and Degradation

Habitat degradation is caused by a number of human activities. Marine mammals that live in coastal environments are most likely to be affected by habitat degradation and loss. Developments such as sewage marine outfalls, moorings, dredging, blasting, dumping, port construction, hydro-electric projects, and aquaculture both degrade the environment and take up valuable habitat. For example, extensive shellfish aquaculture takes up valuable space used by coastal marine mammals for important activities such as breeding, foraging and resting.

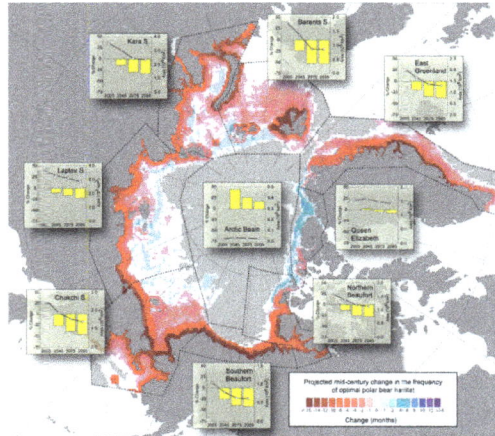

Map from the U.S. Geological Survey shows projected changes in polar bear habitat from 2005 to 2095. Red areas indicate loss of optimal polar bear habitat; blue areas indicate gain.

Contaminants that are discharged into the marine environment accumulate in the bodies of marine mammals when they are stored unintentionally in their blubber along with energy. Contaminants that are found in the tissues of marine mammals include heavy metals, such as mercury and lead, but also organochlorides and polycyclic aromatic hydrocarbons. For example, these can cause disruptive effects on endocrine systems; impair the reproductive system, and lower the immune system of individuals, leading to a higher number of deaths. Other pollutants such as oil, plastic debris and sewage threaten the livelihood of marine mammals.

Noise pollution from anthropogenic activities is another major concern for marine mammals. This is a problem because underwater noise pollution interferes with the abilities of some marine mammals to communicate, and locate both predators and prey. Underwater explosions are used for a variety of purposes including military activities, construction and oceanographic or geophysical research. They can cause injuries such as hemorrhaging of the lungs, and contusion and ulceration of the gastrointestinal tract. Underwater noise is generated from shipping, the oil and gas industry, research, and military use of sonar and oceanographic acoustic experimentation. Acoustic harassment devices and acoustic deterrent devices used by aquaculture facilities to scare away marine mammals emit loud and noxious underwater sounds.

Two changes to the global atmosphere due to anthropogenic activity threaten marine mammals. The first is increases in ultraviolet radiation due to ozone depletion, and this mainly affects the Antarctic and other areas of the southern hemisphere. An increase in ultraviolet radiation has the capacity to decrease phytoplankton abundance, which forms the basis of the food chain in the ocean. The second effect of global climate change is global warming due to increased carbon dioxide levels in the atmosphere. Raised sea levels, sea temperature and changed currents are expected to affect marine mammals by altering the distribution of important prey species, and changing the suitability of breeding sites and migratory routes. The Arctic food chain would be disrupted by the near extinction or migration of polar bears. Arctic sea ice is the polar bear's habitat. It has been declining at a rate of 13% per decade because the temperature is rising at twice the rate of the rest of the world. By the year 2050, up to two-thirds of the world's polar bears may vanish if the sea ice continues to melt at its current rate.

Protection

Logo for the International Whaling Commission

The Marine Mammal Protection Act of 1972 (MMPA) was passed on October 21, 1972 under president Richard Nixon to prevent the further depletion and possible extinction of marine mammal stocks. It prohibits the taking ("the act of hunting, killing, capture, and/or harassment of any marine mammal; or, the attempt at such") of any marine mammal without a permit issued by the Secretary. Authority to manage the MMPA was divided between the Secretary of the Interior through the U.S. Fish and Wildlife Service (Service), and the Secretary of Commerce, which is delegated to the National Oceanic and Atmospheric Administration (NOAA). The Marine Mammal Commission (MMC) was established to review existing policies and make recommendations to the Service and NOAA to better implement the MMPA. The Service is responsible for ensuring the protection of sea otters and marine otters, walruses, polar bears, the three species of manatees, and dugongs; and NOAA was given responsibility to conserve and manage pinnipeds (excluding walruses) and cetaceans.

The 1979 Convention on the Conservation of Migratory Species of Wild Animals (CMS) is the only global organization that conserves a broad range of animals, of which includes marine mammals. Of the agreements made, three of them deal with the conservation of marine mammals: ACCOBAMS, ASCOBANS, and the Wadden Sea Agreement. In 1982, the United Nations Convention on the Law of the Sea (LOSC) adopted a pollution prevention approach to conservation, which many other conventions at the time also adopted.

The Agreement on the Conservation of Cetaceans in the Black Sea, Mediterranean Sea and contiguous Atlantic area (ACCOBAMS), founded in 1996, specifically protects cetaceans in the Mediterranean area, and "maintains a favorable status", a direct action against whaling. There are 23 member states. The Agreement on the Conservation of Small Cetaceans of the Baltic and North Seas (ASCOBANS) was adopted alongside ACCOBAMS to establish a special protection area for Europe's increasingly threatened cetaceans. Other anti-whaling efforts include a ten-year moratorium in 1986 by the IWC on all whaling, and an environmental agreement (a type of international law) the International Convention for the Regulation of Whaling which controlled commercial, scientific, and subsistence whaling.

The Agreement on the Conservation of Seals in the Wadden Sea, enforced in 1991, prohibits the killing or harassment of seals in the Wadden Sea, specifically targeting the harbor seal population.

Various non-governmental organizations participate in marine conservation activism, wherein they draw attention to and aid in various problems in marine conservation, such as pollution, whaling, bycatch, and so forth. Notable organizations include the Greenpeace who focus on over-fishing and whaling among other things, and Sea Shepherd Conservation Society who are known for taking direct-action tactics to expose illegal activity.

As Food

Pilot whale meat (bottom), blubber (middle) and dried fish (left) with potatoes, Faroe Islands

For thousands of years, indigenous peoples of the Arctic have depended on whale meat. The meat is harvested from legal, non-commercial hunts that occur twice a year in the spring and autumn. The meat is stored and eaten throughout the winter. The skin and blubber (muktuk) taken from the bowhead, beluga, or narwhal is also valued, and is eaten raw or cooked. Whaling has also been practiced in the Faroe Islands in the North Atlantic since about the time of the first Norse settlements on the islands. Around 1000 Long-finned pilot whales are still killed annually, mainly during the summer. Today, dolphin meat is consumed in a small number of countries world-wide, which include Japan and Peru (where it is referred to as *chancho marino*, or "sea pork"). In some parts of the world, such as Taiji, Japan and the Faroe Islands, dolphins are traditionally considered food, and are killed in harpoon or drive hunts.

There have been human health concerns associated with the consumption of dolphin meat in Japan after tests showed that dolphin meat contained high levels of methylmercury. There are no known cases of mercury poisoning as a result of consuming dolphin meat, though the government continues to monitor people in areas where dolphin meat consumption is high. The Japanese government recommends that children and pregnant women avoid eating dolphin meat on a regular basis. Similar concerns exist with the consumption of dolphin meat in the Faroe Islands, where prenatal exposure to methylmercury and PCBs primarily from the consumption of pilot whale meat has resulted in neuropsychological deficits amongst children.

The Faroe Islands population was exposed to methylmercury largely from contaminated pilot whale meat, which contained very high levels of about 2 mg methylmercury/kg. However, the Faroe Islands populations also eat significant amounts of fish. The study of about 900 Faroese children showed that prenatal exposure to methylmercury resulted in neuropsychological deficits at 7 years of age

— World Health Organization

Ringed seals were once the main food staple for the Inuit. They are still an important food source for the people of Nunavut and are also hunted and eaten in Alaska. Seal meat is an important source of food for residents of small coastal communities. The seal blubber is used to make seal oil, which is marketed as a fish oil supplement. In 2001, two percent of Canada's raw seal oil was processed and sold in Canadian health stores.

Cetaceans

Performing killer whale at SeaWorld San Diego, 2009

Various species of dolphins are kept in captivity. These small cetaceans are more often than not kept in theme parks and dolphinariums, such as SeaWorld. Bottlenose dolphins are the most common species of dolphin kept in dolphinariums as they are relatively easy to train and have a long lifespan in captivity. Hundreds of bottlenose dolphins live in captivity across the world, though exact numbers are hard to determine. The dolphin "smile" makes them popular attractions, as this is a welcoming facial expression in humans; however the smile is due to a lack of facial muscles and subsequent lack of facial expressions.

Organizations such as World Animal Protection and the Whale and Dolphin Conservation Society campaign against the practice of keeping cetaceans, particularly killer whales, in captivity. In captivity, they often develop pathologies, such as the dorsal fin collapse seen in 60–90% of male killer whales. Captives have vastly reduced life expectancies, on average only living into their 20s. In the wild, females who survive infancy live 46 years on average, and up to 70–80 years in rare cases. Wild males who survive infancy live 31 years on average, and up to 50–60 years. Captivity usually bears little resemblance to wild habitat, and captive whales' social groups are foreign to those found in the wild. Captive life is also stressful due the requirement to perform circus tricks that are not part of wild killer whale behavior, as well as restricting pool size. Wild killer whales may travel up to 100 miles (160 km) in a day, and critics say the animals are too big and intelligent to be suitable for captivity. Captives occasionally act aggressively towards themselves, their tankmates, or humans, which critics say is a result of stress. Dolphins are often trained to do several anthropomorphic behaviors, including waving and kissing—behaviors wild dolphins would rarely do.

Pinnipeds

A sea lion trained to balance a ball on its nose

The large size and playfulness of pinnipeds make them popular attractions. Some exhibits have rocky backgrounds with artificial haul-out sites and a pool, while others have pens with small rocky, elevated shelters where the animals can dive into their pools. More elaborate exhibits contain deep pools that can be viewed underwater with rock-mimicking cement as haul-out areas. The most common pinniped species kept in captivity is the California sea lion as it is abundant and easy to train. These animals are used to perform tricks and entertain visitors. Other species popularly kept in captivity include the grey seal and harbor seal. Larger animals like walruses and Steller sea lions are much less common. Pinnipeds are popular attractions because they are "disneyfied", and, consequently, people often anthropomorphize them with a curious, funny, or playful nature.

Some organizations, such as the Humane Society of the United States and World Animal Protection, object to keeping pinnipeds and other marine mammals in captivity. They state that the exhibits could not be large enough to house animals that have evolved to be migratory, and a pool could never replace the size and biodiversity of the ocean. They also oppose using sea lions for entertainment, claiming the tricks performed are "exaggerated variations of their natural behaviors" and distract the audience from the animal's unnatural environment.

Others

Sea otters can do well in captivity, and are featured in over 40 public aquariums and zoos. The Seattle Aquarium became the first institution to raise sea otters from conception to adulthood with the birth of Tichuk in 1979, followed by three more pups in the early 1980s. In 2007, a YouTube video of two cute sea otters holding paws drew 1.5 million viewers in two weeks, and had over 20 million views as of January 2015. Filmed five years previously at the Vancouver Aquarium, it was

YouTube's most popular animal video at the time, although it has since been surpassed. Otters are often viewed as having a "happy family life", but this is an anthropomorphism.

The oldest manatee in captivity is Snooty, at the South Florida Museum's Parker Manatee Aquarium in Bradenton, Florida. Born at the Miami Aquarium and Tackle Company on July 21, 1948, Snooty was one of the first recorded captive manatee births. He was raised entirely in captivity, and will never be released into the wild. Manatees can also be viewed in a number of European zoos, such as the Tierpark in Berlin, the Nuremberg Zoo, in ZooParc de Beauval in France, and in the Aquarium of Genoa in Italy. The River Safari at Singapore features seven of them.

Military

A dolphin wearing a locating pinger, performing mine clearance work in the Iraq War

Bottlenose dolphins and California sea lions were used in the United States Navy Marine Mammal Program (NMMP) to detect mines, protect ships from enemy soldiers, and recover objects. The Navy has never trained attack dolphins, as they would not be able to discern allied soldiers from enemy soldiers. There were five marine mammal teams, each purposed for one of the three tasks: MK4 (dolphins), MK5 (sea lions), MK6 (dolphins and sea lions), MK7 (dolphins), and MK8 (dolphins); MK is short for mark. The dolphin teams were trained to detect and mark mines either attached to the seafloor or floating in the water column, because dolphins can use their echolocative abilities to detect mines. The sea lion team retrieved test equipment such as fake mines or bombs dropped from planes usually out of reach of divers who would have to make multiple dives. MK6 protects harbors and ships from enemy divers, and was operational in the Gulf War and Vietnam War. The dolphins would swim up behind enemy divers and attach a buoy to their air tank, so that they would float to the surface and alert nearby Navy personnel. Sea lions would hand-cuff the enemy, and try to outmaneuver their counter-attacks.

The use of marine mammals by the Navy, even in accordance with the Navy's policy, continues to meet opposition. The Navy's policy says that only positive reinforcement is to be used while training the military dolphins, and that they be cared for in accordance with accepted standards in animal care. The inevitable stresses involved in training are topics of controversy, as their treatment is unlike the animals' natural lifestyle, especially towards their confined spaces when not training. There is also controversy over the use of muzzles and other inhibitors, which prevent the dolphins from foraging for food while working. The Navy states that this is to prevent them from ingesting harmful objects, but conservation activists say this is done to reinforce the trainers' control over the dolphins, who hand out food rewards. The means of transportation is also an issue for conservation activists, since they are hauled in dry carriers, and switching tanks and introducing the dolphin to new dolphins is potentially dangerous as they are territorial.

References

- Jefferson, T. A.; Leatherwood, S.; Webber, M. A. (1994). Marine Mammals of the World. Food and Agriculture Department of the United Nations. pp. 1–2. ISBN 978-92-5-103292-3. OCLC 30643250.

- Love, John A. (1992). Sea Otters. Golden, Colorado: Fulcrum Publishing. pp. 4–16. ISBN 978-1-55591-123-2. OCLC 25747993.

- Prins, Herbert H. T.; Gordon, Iain J., eds. (2014). "The Biological Invasion of Sirenia into Australasia". Invasion Biology and Ecological Theory. Cambridge: Cambridge University Press. p. 123. ISBN 978-1-107-03581-2. OCLC 850909221.

- Berta, A; Sumich, J. L. (1999). "Exploitation and conservation". Marine Mammals: Evolutionary Biology. San Diego: Academic Press. ISBN 978-0-12-093225-2. OCLC 42467530.

- Riedman, M. (1990). The Pinnipeds: Seals, Sea Lions, and Walruses. Los Angeles: University of California Press. ISBN 978-0-520-06497-3. OCLC 19511610.

- Whitehead, H. (2003). Sperm Whales: Social Evolution in the Ocean. Chicago: University of Chicago Press. p. 79. ISBN 978-0-226-89518-5. OCLC 51242162.

- Silverstein, Alvin; Silverstein, Virginia; Silverstein, Robert (1995). The Sea Otter. Brookfield, Connecticut: The Millbrook Press, Inc. p. 19. ISBN 978-1-56294-418-6. OCLC 30436543.

- Kenyon, Karl W. (1975). The Sea Otter in the Eastern Pacific Ocean. New York: Dover Publications. ISBN 978-0-486-21346-0. OCLC 1504461.

- Lockyer, C. J. H.; Brown, S. G. (1981). "The Migration of Whales". In Aidley, D. Animal Migration. CUP Archive. p. 111. ISBN 978-0-521-23274-6.

- Klinowska, Margaret; Cooke, Justin (1991). Dolphins, Porpoises, and Whales of the World: the IUCN Red Data Book (PDF). Columbia University Press, NY: IUCN Publications. ISBN 978-2-88032-936-5. OCLC 24110680.

- Berta, A.; Sumich, J. L.; Kovacs, K. M. (2015). Marine Mammals: Evolutionary Biology. London: Academic Press. p. 430. ISBN 978-0-12-397002-2. OCLC 905649783.

- Haley, D., ed. (1986). "Sea Otter". Marine Mammals of Eastern North Pacific and Arctic Waters (2nd ed.). Seattle, Washington: Pacific Search Press. ISBN 978-0-931397-14-1. OCLC 13760343.

- VanBlaricom, Glenn R. (2001). Sea Otters. Stillwater, MN: Voyageur Press Inc. pp. 22, 33, 69. ISBN 978-0-89658-562-1. OCLC 46393741.

- Lavinge, D. M.; Kovacs, K. M.; Bonner, W. N. (2001). "Seals and Sea lions". In MacDonald, D. The Encyclopedia of Mammals (2nd ed.). Oxford University Press. pp. 147–55. ISBN 978-0-7607-1969-5. OCLC 48048972.

Understanding Aquatic Mammals

Aquatic mammals are animals that live partly or entirely in water bodies. They include animals such as the European otter and the Amazon river dolphin. Blubber, cetacean bycatch, aquatic locomotion, marine mammal observer and marine mammals and sonar are some aspects of aquatic mammals. This section is an overview of the subject matter incorporating all the major aspects of aquatic mammal.

Aquatic Mammal

Aquatic and semiaquatic mammals are a diverse group of mammals that dwell partly or entirely in bodies of water. They include the various marine mammals who dwell in oceans, as well as various freshwater species, such as the European otter. They are not a taxon and are not unified by any distinct biological grouping, but rather their dependence on and integral relation to aquatic ecosystems. The level of dependence on aquatic life are vastly different among species, with the Amazonian manatee and river dolphins being completely aquatic and fully dependent on aquatic ecosystems; whereas the Baikal seal feeds underwater but rests, molts, and breeds on land; and the capybara and hippopotamus are able to venture in and out of water in search of food.

An Amazon river dolphin (*Inia geoffrensis*), a member of the infraorder Cetacea of the order Cetartiodactyla

Mammal adaptation to an aquatic lifestyle vary considerably between species. River dolphins and manatees are both fully aquatic and therefore are completely tethered to a life in the water. Seals are semiaquatic; they spend the majority of their time in the water, but need to return to land for important activities such as mating, breeding and molting. In contrast, many other aquatic mammals, such as rhinoceroses, capybara, and water shrews, are much less adapted to aquatic living. Likewise, their diet ranges considerably as well, anywhere from aquatic plants and leaves to small fish and crustaceans. They play major roles in maintaining aquatic ecosystems, beavers especially.

A Ladoga seal (*Pusa hispida ladogensis*), a member of the clade Pinnipedia of the order Carnivora

Aquatic mammals were the target for commercial industry, leading to a sharp decline in all populations of exploited species, such as beavers. Their pelts, suited for conserving heat, were taken during the fur trade and made into coats and hats. Other aquatic mammals, such as the Indian rhinoceros, were targets for sport hunting and had a sharp population decline in the 1900s. After it was made illegal, many aquatic mammals became subject to poaching. Other than hunting, aquatic mammals can be killed as bycatch from fisheries, where they become entangled in fixed netting and drown or starve. Increased river traffic, most notably in the Yangtze river, causes collisions between fast ocean vessels and aquatic mammals, and damming of rivers may land migratory aquatic mammals in unsuitable areas or destroy habitat upstream. The industrialization of rivers led to the extinction of the Chinese river dolphin, with the last confirmed sighting in 2004.

Taxonomy and Evolution

Aquatic mammals vary greatly in size and shape		
A North American beaver (*Castor canadensis*)	Eurasian otter (*Lutra lutra*) in Southwold, Suffolk, England	A partially submerged Indian rhinoceros (*Rhinoceros unicornis*) at the Cincinnati Zoo and Botanical Garden

Groups

This list covers only mammals that live in freshwater.

- **Order Sirenia**: sirenians
  - Family *Trichechidae*: manatees
    - Amazonian manatee (*Trichechus inunguis*)
    - African manatee (*Trichechus senegalensis*)
    - Dwarf manatee (*Trichechus pygmaeus*) validity questionable

- **Order Cetartiodactyla**: even-toed ungulates
  - *Suborder Whippomorpha*
    - Family Platanistidae
      - South Asian river dolphin (*Platanista gangetica*) with two subspecies
        - Ganges river dolphin, or susu (*Platanista gangetica gangetica*)
        - Indus river dolphin, or bhulan (*Platanista gangetica minor*)
    - Family Iniidae
      - Amazon river dolphin, or boto (*Inia geoffrensis*)
      - Araguaian river dolphin (*Inia araguaiaensis*)
    - Family Lipotidae
      - Chinese river dolphin, or baiji (*Lipotes vexillifer*) functionally extinct since December 2006
    - Family Pontoporiidae
      - La Plata dolphin, or franciscana (*Pontoporia blainvillei*)
    - Family Hippopotamidae: hippopotamuses
      - Hippopotamus (*Hippopotamus amphibius*)
      - Pygmy hippopotamus (*Choeropsis liberiensis*)
  - *Suborder Ruminantia*
    - Family Cervidae
      - Moose (*Alces alces*)

- **Order Carnivora**
  - **Family Mustelidae**
    - *Subfamily Lutrinae*

- Eurasian otter (*Lutra lutra*)

- Hairy-nosed otter (*Lutra sumatrana*)

- Spotted-necked otter (*Hydrictis maculicollis*)

- Smooth-coated otter (*Lutrogale perspicillata*)

- North American river otter (*Lontra canadensis*)

- Southern river otter (*Lontra provocax*)

- Neotropical river otter (*Lontra longicaudis*)

- Giant otter (*Pteronura brasiliensis*)

- African clawless otter (*Aonyx capensis*)

- Oriental small-clawed otter (*Aonyx cinerea*)

- *Subfamily Mustelinae*

- European mink (*Mustela lutreola*)

- American mink (*Neovison vison*)

o **Family Phocidae**

- Genus *Pusa*

- Baikal seal (*Pusa sibirica*)

- Ladoga seal (*Pusa hispida ladogensis*)

- Saimaa seal (*Pusa hispida saimensis*)

- **Order Rodentia**: rodents

o Suborder Hystricomorpha

- Capybara (*Hydrochoerus hydrochaeris*)

- Lesser capybara (*Hydrochoerus isthmius*)

- Coypu (*Myocastor coypus*)

o Family Castoridae: beavers

- North American beaver (*Castor canadensis*)

- Eurasian beaver (*Castor fiber*)

o Family Cricetidae

- Muskrat (*Ondatra zibethicus*)

- European water vole (*Arvicola amphibius*)

- **Order Monotremata**: monotremes
  - Platypus (Ornithorhynchus anatinus)

- **Order Perissodactyla**:odd-toed ungulates
  - Family Rhinocerotidae: rhinoceroses
    - Indian rhinoceros (*Rhinoceros unicornis*)

- **Order Afrosoricida**
  - Giant otter shrew (*Potamogale velox*)

- **Order Soricomorpha**
  - Family Soricidae: shrews
    - Malayan water shrew (*Chimarrogale hantu*)
    - Himalayan water shrew (*Chimarrogale himalayica*)
    - Sunda water shrew (*Chimarrogale phaeura*)
    - Japanese water shrew (*Chimarrogale platycephala*)
    - Chinese water shrew (*Chimarrogale styani*)
    - Sumatran water shrew (*Chimarrogale sumatrana*)
    - Elegant water shrew (*Nectogale elegans*)
    - Mediterranean water shrew (*Neomys anomalus*)
    - Eurasian water shrew (*Neomoys fodiens*)
    - Transcaucasian water shrew (*Neomys teres*)
    - Glacier Bay water shrew (*Sorex alaskanus*)
    - American water shrew (*Sorex palustris*)
    - Pacific water shrew, or marsh shrew (*Sorex bendirii*)
  - Family Talpidae (moles and relatives)
    - Russian desman (*Desmana moschata*)

- **Order Didelphimorphia:** opossums
  - Family Didelphidae: opossums
    - Lutrine opossum (*Lutreolina crassicaudata*)
    - Yapok (*Chironectes minimus*)

Evolution

Mesozoic

The teeth of *Castorocauda*, a presumably beaver-like mammal, are different in many ways from all other docodonts, presumably due to a difference in diet. Most docodonts had teeth specialized for an omnivorous diet. The teeth of *Castorocauda* suggest that the animal was a piscivore, feeding on fish and small invertebrates. The first two molars had cusps in a straight row, eliminating the grinding function suggesting that they were strictly for gripping and not for chewing. This feature of three cusps in a row is similar to the ancestral condition in mammal relatives (as seen in triconodonts), but is almost certainly a derived character in *Castorocauda*. These first molars were also recurved in a manner designed to hold slippery prey once grasped. These teeth are very similar to the teeth seen in mesonychids, an extinct group of semi-aquatic carnivorous ungulates, and resemble, to a lesser degree, the teeth of seals.

Illustration of *Castorocauda lutrasimilis*, a semi-aquatic mammal from the Jurassic

Another docodontan, the Late Jurassic *Haldanodon*, has been suggested to be a platypus or desman-like swimmer and burrower, being well adapted to dig and swim and occurring in a wetland environment.

The tritylodontid *Kayentatherium* has been suggested to be semi-aquatic. Unlike *Castorocauda* and *Haldanodon*, it was a herbivore, being probably beaver or capybara-like in habits.

Another lineage of Mesozoic mammals, the eutriconodonts, have been suggested to be aquatic animals with mixed results. *Astroconodon* occurred abundantly in freshwater lacustrine deposits and its molars were originally interpreted as being similar to those of piscivorous mammals like cetaceans and pinnipeds; by extension some researchers considered the possibility that all eutriconodonts were aquatic piscivores. However, Zofia Kielan-Jaworowska and other researchers have latter found that the triconodont molars of eutriconodonts were more suited for a carnassial-like shearing action than the piercing and gripping function of piscivorous mammal molars, occluding instead of interlocking, and that *Astroconodon*'s aquatic occurrences may be of little significance when most terrestrial tetrapod fossils are found in lacustrine environments anyway.

However, two other eutriconodonts, *Dyskritodon* and *Ichthyoconodon*, occur in marine deposits with virtually no dental erosion, implying that they died *in situ* and are thus truly aquatic mammals. Nonetheless, *Ichthyoconodon* may not be aquatic, but instead a gliding or even flying mammal. More recently, *Yanoconodon* and *Liaoconodon* have been interpreted as semi-aquatic, bearing a long body and paddle-like limbs.

Life restoration of the eutriconodont *Liaoconodon*.

A metatherian, the stagodontid *Didelphodon*, has been suggested to be aquatic, due to molar and skeleton similarities to sea otters.

Cenozoic

An extinct genus, *Satherium*, is believed to be ancestral to South American river otters, having migrated to the New World during the Pliocene or early Pleistocene. The South American continent houses the *Lontra* genus of otters: the giant otter, the neotropical river otter, the southern river otter, and the marine otter. The smooth-coated otter (*Lutrogale perspicillata*) of Asia may be its closest extant relative; similar behaviour, vocalizations, and skull morphology have been noted.

The most popular theory of the origins of Hippopotamidae suggests that hippos and whales shared a common ancestor that branched off from other artiodactyls around 60 million years ago (mya). This hypothesized ancestral group likely split into two branches around 54 mya. One branch would evolve into cetaceans, possibly beginning about 52 mya, with the protowhale *Pakicetus* and other early whale ancestors collectively known as Archaeoceti, which eventually underwent aquatic adaptation into the completely aquatic cetaceans. The other branch became the anthracotheres, and all branches of the anthracotheres, except that which evolved into Hippopotamidae, became extinct during the Pliocene without leaving any descendants. River dolphins are thought to have relictual distributions, that is, their ancestors originally occupied marine habitats, but were then displaced from these habitats by modern dolphin lineages. Many of the morphological similarities and adaptations to freshwater habitats arose due to convergent evolution; thus, a grouping of all river dolphins is paraphyletic. For example, Amazon river dolphins are actually more closely related to oceanic dolphins than to South Asian river dolphins.

Illustration of *Prorastomus*, a sirenian from the Eocene

Sirenians, along with Proboscidea (elephants), group together with the extinct Desmostylia and likely the extinct Embrithopoda to form the Tethytheria. Tethytheria is thought to have evolved from primitive hoofed mammals ("condylarths") along the shores of the ancient Tethys Ocean. Tethytheria, combined with Hyracoidea (hyraxes), forms a clade called Paenungulata. Paenungulata and Tethytheria (especially the latter) are among the least controversial mammalian clades, with strong support from morphological and molecular interpretations. That is, elephants, hyraxes, and manatees share a common ancestry. The ancestry of Sirenia is distinct from that of Cetacea and Pinnipedia, although they are thought to have evolved an aquatic lifestyle around the same time.

The oldest fossil of the modern platypus dates back to about 100,000 years ago, during the Quaternary period. The extinct monotremes *Teinolophos* and *Steropodon* were once thought to be closely related to the modern platypus, but more recent studies show that platypi are more related to the modern echidnas than to these ancient forms and that at least *Teinolophos* was a rather different mammal lacking several speciations seen in platypi. However, the last common ancestor between platypi and echidnas probably was aquatic, and echidnas thus secondarily became terrestrial. *Monotrematum sudamericanum* is currently the oldest aquatic monotreme known. It has been found in Argentina, indicating monotremes were present in the supercontinent of Gondwana when the continents of South America and Australia were joined via Antarctica, or that monotremes existed along the shorelines of Antarctica in the early Cenozoic.

Marine Mammals

The humpback whale is a fully aquatic marine mammal

Marine mammals are aquatic mammals that rely on the ocean for their existence. They include animals such as sea lions, whales, dugongs, sea otters and polar bears. Like other aquatic mammals, they do not represent a biological grouping.

Marine mammal adaptation to an aquatic lifestyle vary considerably between species. Both cetaceans and sirenians are fully aquatic and therefore are obligate ocean dwellers. Pinnipeds are semiaquatic; they spend the majority of their time in the water, but need to return to land for important activities such as mating, breeding and molting. In contrast, both otters and the polar bear are much less adapted to aquatic living. Likewise, their diet ranges considerably as well; some may eat zooplankton, others may eat small fish, and a few may eat other mammals. While the number of marine mammals is small compared to those found on land, their roles in various ecosystems are large. They, namely sea otters and polar bears, play important roles in maintaining marine ecosystems, especially through regulation of prey populations. Their role in maintaining ecosystems makes them of particular concern considering 23% of marine mammal species are currently threatened.

Marine mammals were first hunted by aboriginal peoples for food and other resources. They were also the target for commercial industry, leading to a sharp decline in all populations of exploited species, such as whales and seals. Commercial hunting lead to the extinction of Steller's sea cow and the Caribbean monk seal. After commercial hunting ended, some species, such as the gray whale and northern elephant seal, have rebounded in numbers, however the northern elephant seal has a genetic bottleneck; conversely, other species, such as the North Atlantic right whale, are critically endangered. Other then hunting, marine mammals, dolphins especially, can be killed as bycatch from fisheries, where they become entangled in fixed netting and drown or starve. Increased ocean traffic causes collisions between fast ocean vessels and large marine mammals. Habitat degradation also threatens marine mammals and their ability to find and catch food. Noise pollution, for example, may adversely affect echolocating mammals, and the ongoing effects of global warming degrades arctic environments.

Adaptations

Mammals evolved on land, so all aquatic and semiaquatic mammals have brought many terrestrial adaptations into the waters. They do not breathe underwater as fish do, so their respiratory systems had to protect the body from the surrounding water; valvular nostrils and an intranarial larynx exclude water while breathing and swallowing. To navigate and detect prey in murky and turbid waters, aquatic mammals have developed a variety of sensory organs: for example, manatees have elongated and highly sensitive whiskers which are used to detect food and other vegetation directly front of them, and toothed whales have evolved echolocation.

The pygmy hippopotamus has four weight-bearing limbs, and can walk on land like a fully terrestrial mammal.

Aquatic mammals also display a variety of locomotion styles. Cetaceans excel in streamlined body shape and the up-and-down movements of their flukes make them fast swimmers; the tucuxi, for example, can reach speeds of 14 miles per hour (23 km/h). The considerably slower sirenians can also propel themselves with their fluke, but they can also walk on the bottom with their forelimbs.

The earless seals (Phocidae) swim by moving their hind-flippers and lower body from side to side, while their fore-flippers are mainly used for steering. They are clumsy on land, and move on land by lunging, bouncing and wiggling while their fore-flippers keep them balanced; when confronted with predators, they retreat to the water as freshwater phocids have no aquatic predators.

Some aquatic mammals have retained four weight-bearing limbs (e.g. hippopotamuses, beavers, otters, muskrats) and can walk on land like fully terrestrial mammals. The long and thin legs of a moose limit exposure to and friction from water in contrast to hippopotamuses who keep most of their body submerged and have short and thick legs. The semiaquatic pygmy hippopotamus can walk quickly on a muddy underwater surface thanks to robust muscles and because all toes are weight-bearing. Some aquatic mammals with flippers (e.g. seals) are amphibious and regularly leave the water, sometimes for extended periods, and maneuver on land by undulating their bodies to move on land, similar to the up-and-down body motion used underwater by fully aquatic mammals (e.g. dolphins and manatees).

Beavers, muskrats, otters, capybara have fur, one of the defining mammalian features, that is long, oily, and waterproof in order to trap air to provide insulation. In contrast, other aquatic mammals, such as dolphins, manatees, seals, and hippopotamuses, have lost their fur in favor of a thick and dense epidermis, and a thickened fat layer (blubber) in response to hydrodynamic requirements.

Wading and bottom-feeding animals (e.g. moose and manatee) need to be heavier than water in order to keep contact with the floor or to stay submerged, surface-living animals (e.g. otters) need the opposite, and free-swimming animals living in open waters (e.g. dolphins) need to be neutrally buoyant in order to be able to swim up and down the water column. Typically, thick and dense bone is found in bottom feeders and low bone density is associated with mammals living in deep water.

The shape and function of the eyes in aquatic animals are dependent on water depth and light exposure: limited light exposure results in a retina similar to that of nocturnal terrestrial mammals. Additionally, cetaceans have two areas of high ganglion cell concentration ("best-vision areas"), where other aquatic mammals (e.g. seals, manatees, otters) only have one.

Among non-placental mammals, which cannot give birth to fully developed young, some adjustments have been made for an aquatic lifestyle. The yapok has a backwards-facing pouch which seals off completely when the animal is underwater, while the platypus deposits its young on a burrow on land.

Ecology

Keystone Species

Beaver ponds have a profound effect on the surrounding ecosystem. Their first and foremost ecological function is as a reservoir for times of drought, and prevent drying of riverbeds. In the event of a flood, beaver ponds slow down water-flow which reduces erosion on the surrounding soil. Beaver dams hold sediment, which reduces turbidity and thereby improving overall water quality downstream. This supplies other animals with cleaner drinking water and prevents degradation of spawning grounds for fish. However, the slower water speed and lack of shade from trees (that have since been cut down to construct the dam), the overall temperature increases. They also house predatory zooplankton which help break down detritus and control algae populations.

Beaver dams restrict water-flow, creating a pond

Diet

Beavers are herbivores, and prefer the wood of quaking aspen, cottonwood, willow, alder, birch, maple and cherry trees. They also eat sedges, pondweed, and water lilies. Beavers do not hibernate, but rather they store sticks and logs in a pile in their ponds, eating the underbark. The dams they build flood areas of surrounding forest, giving the beaver safe access to an important food supply, which is the leaves, buds, and inner bark of growing trees. They prefer aspen and poplar, but will also take birch, maple, willow, alder, black cherry, red oak, beech, ash, hornbeam and occasionally pine and spruce. They will also eat cattails, water lilies and other aquatic vegetation, especially in the early spring.

Indian rhinoceros are grazers. Their diets consist almost entirely of grasses, but they also eat leaves, branches of shrubs and trees, fruits, and submerged and floating aquatic plants.They feed in the mornings and evenings. They use their prehensile lips to grasp grass stems, bend the stem down, bite off the top, and then eat the grass. They tackle very tall grasses or saplings by walking over the plant, with legs on both sides and using the weight of their bodies to push the end of the plant down to the level of the mouth.

Manatees make seasonal movements synchronized with the flood regime of the Amazon Basin. They are found in flooded forests and meadows during the flood season when food is abundant, and move to deep lakes during the dry season. The Amazonian manatee has the smallest degree of rostral deflection (25° to 41°) among sirenians, an adaptation to feed closer to the water surface.

A moose's diet often depends on its location, but they seem to prefer the new growths from deciduous trees with a high sugar content, such as white birch, trembling aspen and striped maple, among many others. They also eat many aquatic plants such as lilies and water milfoil. To reach high branches, a moose may bend small saplings down, using its prehensile lip, mouth or body. For larger trees a moose may stand erect and walk upright on its hind legs, allowing it to reach plants 14.0 feet (4.26 m) off the ground. Moose are excellent swimmers and are known to wade into water to eat aquatic plants. Moose are thus attracted to marshes and river banks during warmer months as both provide suitable vegetation to eat and water to bathe in. Moose have been known to dive underwater to reach plants on lake bottoms, and the complex snout may assist the moose in this type of feeding. Moose are the only deer that are capable of feeding underwater.

A moose in Siberia feeding on aquatic plants

Hippopotamuses leave the water at dusk and travel inland, sometimes up to 10 km (6 mi), to graze on short grasses, their main source of food. They spend four to five hours grazing and can consume 68 kg (150 lb) of grass each night. Like almost any herbivore, they consume other plants if presented with them, but their diet consists almost entirely of grass, with only minimal consumption of aquatic plants. The pygmy hippopotamus emerges from the water at dusk to feed. It relies on game trails to travel through dense forest vegetation. It marks trails by vigorously waving its tail while defecating to further spread its feces. The pygmy hippo spends about six hours a day foraging for food, and they do not eat aquatic vegetation to a significant extent and rarely eat grass because it is uncommon in the thick forests they inhabit. The bulk of a pygmy hippo's diet consists of ferns, broad-leaved plants and fruits that have fallen to the forest floor. The wide variety of plants pygmy hippos have been observed eating suggests that they will eat any plants available. This diet is of higher quality than that of the common hippopotamus.

The Amazon river dolphin has the most diverse diet among cetaceans, consisting of at least 53 species of fish. They mainly feed on croakers, cichlids, tetras, and pirahnas, but they may also target freshwater crabs and river turtles. South Asian river dolphins mainly eat fish (such as carp, catfish, and freshwater sharks) and invertebrates, mainly prawns.

Generally, all aquatic desmans, shrews, and voles make quick dives and catch small fish and invertebrates. The giant otter shrew, for example, makes quick dives that last for seconds and grabs small crabs (usually no bigger than 2.8 inches (7 cm) across). The Lutrine opossum is the most carnivorous opossum, usually consuming small birds, rodents, and invertebrates. Water voles mainly eat grass and plants near the water and at times, they will also consume fruits, bulbs, twigs, buds, and roots. However, a population of water voles living in Wiltshire and Lincolnshire, England have started eating frogs' legs and discarding the bodies.

Interactions with Humans

Exploitation

Fur robes were blankets of sewn-together, native-tanned, beaver pelts. The pelts were called *castor gras* in French and "coat beaver" in English, and were soon recognized by the newly developed felt-hat making industry as particularly useful for felting. Some historians, seeking to explain the term *castor gras*, have assumed that coat beaver was rich in human oils from having been worn so long (much of the top-hair was worn away through usage, exposing the valuable under-wool), and that this is what made it attractive to the hatters. This seems unlikely, since grease interferes with the felting of wool, rather than enhancing it. By the 1580s, beaver "wool" was the major starting material of the French felt-hatters. Hat makers began to use it in England soon after, particularly after Huguenot refugees brought their skills and tastes with them from France.

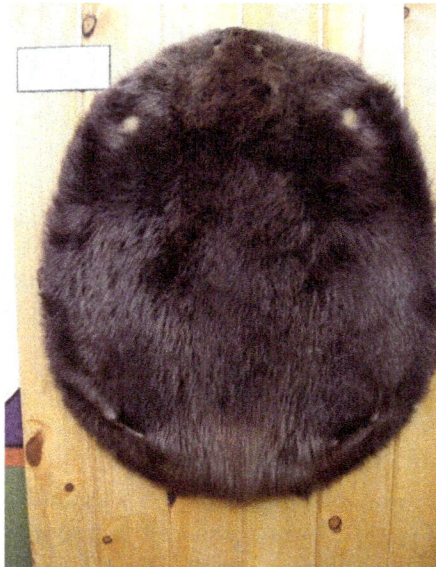

A beaver pelt in the Fur Trade Museum

Sport hunting of the Indian rhinoceros became common in the late 1800s and early 1900s. Indian rhinos were hunted relentlessly and persistently. Reports from the middle of the 19th century claim that some British military officers in Assam individually shot more than 200 rhinos. By 1908, the population in Kaziranga had decreased to around 12 individuals. In the early 1900s, the species had declined to near extinction. Poaching for rhinoceros horn became the single most important reason for the decline of the Indian rhino after conservation measures were put in place from the beginning of the 20th century, when legal hunting ended. From 1980 to 1993, 692 rhinos were poached in India. In India's Laokhowa Wildlife Sanctuary, 41 rhinos were killed in 1983, virtually the entire population of the sanctuary. By the mid-1990s, poaching had rendered the species extinct there. In 1950, Chitwan's forest and grasslands extended over more than 2,600 km² (1,000 sq mi) and were home to about 800 rhinos. When poor farmers from the mid-hills moved to the Chitwan Valley in search of arable land, the area was subsequently opened for settlement, and poaching of wildlife became rampant. The Chitwan population has repeatedly been jeopardized by poaching; in 2002 alone, poachers killed 37 animals to saw off and sell their valuable horns.

Otters have been hunted for their pelts since at least the 1700s. There has been a long history of otter pelts being worn around the world. In China it was standard for the royalty to wear robes made from them. People that were financially high in status also wore them. Otters have also been hunted using dogs, specifically the otterhound. In modern times, TRAFFIC, a joint program of the World Wildlife Fund (WWF) and International Union for Conservation of Nature (IUCN), reported that otters are at serious risk in Southeast Asia and have disappeared from parts of their former range. This decline in populations is due to hunting to supply the demand for skins.

Habitat Degradation

The Baykalsk Pulp and Paper Mill was a major producer of industrial waste for Lake Baikal.

One problem at Lake Baikal is the introduction of pollutants into the ecosystem. Pesticides such as DDT and hexachlorocyclohexane, as well as industrial waste, mainly from the Baykalsk Pulp and Paper Mill, are thought to have been the cause of several disease epidemics among Baikal seal populations. The chemicals are speculated to concentrate up the food chain and weaken the Baikal seal's immune system, making them susceptible to diseases such as canine distemper and the plague, which was the cause of a serious Baikal seal epidemic that resulted in the deaths of thousands of animals in 1997 and 1999. Baikal seal pups have higher levels of DDT and PCB than known in any other population of European or Arctic earless seal.

In the 1940s, beavers were brought from Canada to the island of Tierra Del Fuego in southern Chile and Argentina, for commercial fur production. However, the project failed and the beavers, ten pairs, were released into the wild. Having no natural predators in their new environment, they quickly spread throughout the island, and to other islands in the region, reaching a number of 100,000 individuals within just 50 years. They are now considered a serious invasive species in the region, due to their massive destruction of forest trees, and efforts are being made for their eradication.

In some European countries, such as Belgium, France, and the Netherlands, the muskrat is considered an invasive pest, as its burrowing damages the dikes and levees on which these low-lying countries depend for protection from flooding. In those countries, it is trapped, poisoned, and hunted to attempt to keep the population down. Muskrats also eat corn and other farm and garden crops growing near water bodies.

Urban and agricultural development, increased damming, and increased use of hydroelectric power in rivers in countries such as Côte d'Ivoire and Ghana are threats to the African manatee's habitat and life, and thick congestion of boats in waterways may cause them to have a deadly run-in with the vessels. However, even natural occurrences, such as droughts and tidal changes, often strand manatees in an unsuitable habitat. Some are killed accidentally by fishing trawls and nets intended for catching sharks. The Amazonian manatee is at risk from pollution, accidental drowning in commercial fishing nets, and the degradation of vegetation by soil erosion resulting from deforestation. Additionally, the indiscriminate release of mercury in mining activities threatens the entire aquatic ecosystem of the Amazon Basin.

Dead trees as a result of the construction of a beaver dam in Tierra del Fuego

As China developed economically, pressure on the Chinese river dolphin grew significantly. Industrial and residential waste flowed into the Yangtze. The riverbed was dredged and reinforced with concrete in many locations. Ship traffic multiplied, boats grew in size, and fishermen employed wider and more lethal nets. Noise pollution caused the nearly blind animal to collide with propellers. Stocks of the dolphin's prey declined drastically in the late 20th century, with some fish populations declining to one thousandth of their pre-industrial levels. In the 1950s, the population was estimated at 6,000 animals, but declined rapidly over the subsequent five decades. Only a few hundred were left by 1970. Then the number dropped down to 400 by the 1980s and then to 13 in 1997 when a full-fledged search was conducted. On December 13, 2006, the baiji was declared functionally extinct, after a 45-day search by leading experts in the field failed to find a single specimen. The last verified sighting was in 2004.

As Food

Moose are hunted as a game species in many of the countries where they are found. While the flesh has protein levels similar to those of other comparable red meats (e.g. beef, deer and elk), it has a low fat content, and the fat that is present consists of a higher proportion of polyunsaturated fats rather than saturated fats.

...like tender beef, with perhaps more flavour; sometimes like veal"

— *Henry David Thoreau of The Maine Woods describing the taste of moose meat*

Cadmium levels are high in moose liver and kidneys, with the result that consumption of these organs from moose more than one year old is prohibited in Finland. Cadmium intake has been found to be elevated amongst all consumers of moose meat, though the meat was found to contribute only slightly to the daily cadmium intake. However the consumption of moose liver or kidneys significantly increased cadmium intake, with the study revealing that heavy consumers of moose organs have a relatively narrow safety margin below the levels which would probably cause adverse health effects.

In the 17th century, based on a question raised by the Bishop of Quebec, the Roman Catholic Church ruled that the beaver was a fish (beaver flesh was a part of the indigenous peoples' diet, prior to the Europeans' arrival) for purposes of dietary law. Therefore, the general prohibition on the consumption of meat on Fridays did not apply to beaver meat. This is similar to the Church's classification of other semi-aquatic rodents, such as the capybara and muskrat.

Blubber

Whale blubber

Blubber is a thick layer of vascularized adipose tissue found under the skin of all cetaceans, pinnipeds and sirenians.

Description

Lipid-rich, collagen fiber-laced blubber comprises the hypodermis and covers the whole body, except for parts of the appendages, strongly attached to the musculature and skeleton by highly organized, fan-shaped networks of tendons and ligaments. It can comprise up to 50% of the body mass of some marine mammals during some points in their lives, and can range from 2 inches (5 cm) thick in dolphins and smaller whales, to more than 12 inches (30 cm) thick in some bigger whales, such as right and bowhead whales. However, this is not indicative of larger whales' ability to retain heat better, as the thickness of a whale's blubber does not significantly affect heat loss. More indicative of a whale's ability to retain heat is the water and lipid concentration in blubber, as water reduces heat-retaining capacities, and lipid increases them.

Function

Blubber is the primary storage location of fat on some mammals. It is particularly important for species that feed and breed in different parts of the ocean. During these periods, the animals metabolize fat. Blubber may save energy for marine mammals such as dolphins, in that it adds buoyancy to a dolphin while swimming.

Blubber differs from other forms of adipose tissue in its extra thickness, which allows it to serve as an efficient thermal insulator, making blubber essential for thermoregulation. Blubber is more vascularized—rich in blood vessels—than other adipose tissue.

Blubber has advantages over fur (as in sea otters) in the respect that although fur can retain heat by holding pockets of air, the air pockets will be expelled under pressure (while diving). Blubber, however, does not compress under pressure. It is effective enough that some whales can dwell in temperatures as low as 40 °F (4 °C). While diving in cold water, blood vessels covering the blubber constrict and decrease blood flow, thus increasing blubber's efficiency as an insulator.

Blubber aids buoyancy, and streamlines the body because the organized, complex collagenous network supports the noncircular cross sections characteristic of cetaceans.

Research into the thermal conductivity of the common bottlenose dolphin's blubber reveals its thickness and lipid content vary greatly amongst individuals and across life history categories. Blubber from emaciated dolphins is a poorer insulator than that from nonpregnant adults, which in turn have a higher heat conductivity than blubber from pregnant females and adolescents.

Human Influences

Uses

Uqhuq or *uqsuq*, ("blubber" in the Inuktitut language) formed an important part of the traditional diets of the Inuit and of other northern peoples because of its high energy value and availability. Seal blubber contains large amounts of vitamin E, selenium, and other antioxidants, which may reduce the effect of the free radicals formed within the body's cells. (Damage caused to cells by free radicals is a theorized contributor to oxidative diseases.) Whale blubber, which tastes like arrowroot biscuits, has similar properties. The positive effects of consuming blubber can be seen in Greenland; in Uummannaq for example, a hunting district with 3,000 residents, no deaths due to cardiovascular diseases occurred in the 1970s. However, emigrants to Denmark have contracted the same diseases as the rest of the Danish population. The average 70-year-old Inuit with a traditional diet of whale and seal has arteries as elastic as those of a 20-year-old resident of Denmark.

Whaling largely targeted the collection of blubber: whalers rendered it into oil in try pots, or later, in vats on factory ships. The oil could serve in the manufacture of soap, leather, and cosmetics. Whale oil was used in candles as wax, and in oil lamps as fuel.

A single blue whale can yield a blubber harvest of up to 50 tons.

Health

Blubber from whales and seals contains omega-3 fatty acids and vitamin D. Without the vitamin D, for example, the Inuit and other natives of the Arctic would likely suffer from rickets. There is evidence blubber and other fats in the arctic diet also provide the calories needed to replace the lack of carbohydrates found in the diets of cultures in the rest of the world.

Toxicity

Blubber contains PCBs, carcinogens that damage human nervous, immune and reproductive systems. The source of PCB concentrations is unknown. Since toothed whales are high on the food chain, they likely consume large amounts of industrial pollutants (bioaccumulation). Even baleen whales, by merit of the huge amount of food they consume, are bound to have toxic chemicals stored in their bodies. There are high levels of mercury in the blubber of seals of the Canadian arctic.

Cetacean Bycatch

Group of Fraser's dolphins.

Cetacean bycatch is the incidental capture of non-target cetacean species such as dolphins, porpoises, and whales by fisheries. Bycatch can be caused by entanglement in fishing nets and lines, or direct capture by hooks or in trawl nets.

Cetacean bycatch is increasing in intensity and frequency. This is a trend that is likely to continue because of increasing human population growth and demand for marine food sources, as well as industrialization of fisheries which are expanding into new areas. These fisheries come into direct and indirect contact with cetaceans. An example of direct contact is the physical contact of cetaceans with fishing nets. Indirect contact is through marine trophic pathways where fisheries are severely reducing fish stocks that cetaceans rely on for food. In some fisheries, cetaceans are captured as bycatch but then retained because of their value as food or bait. In this fashion, cetaceans can become a target of fisheries.

Bycatch Trends

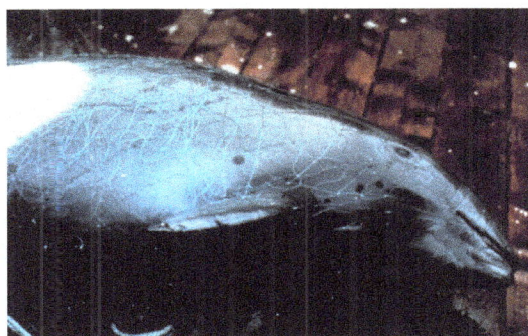
A Dall's Porpoise caught in a fishing net

Generally cetacean bycatch is on the increase. Most of the world's cetacean bycatch occurs in gill-net fisheries. The mean annual bycatch in the U.S. alone from 1990–1999 was 6,215 marine mammals, with dolphins and porpoises being the primary cetaceans caught in gillnets. A study by Read et al. estimated global bycatch through observation of U.S. fisheries and came to the conclusion that an annual estimate of 653,365 marine mammals, comprising 307,753 cetaceans and 345,611 pinnipeds were caught from 1990–1994.

While gillnets are a principal concern, other types of nets also pose a problem: trawl nets, purse seines, beach seines, longline gear, and driftnets. Driftnets are known for high rates of bycatch and they affect all cetaceans and other marine species. They are fatal for small toothed whales (*Odontocetes*) and sperm whales, as well as other marine mammals and fish such as sharks, sea birds and sea turtles. Many fisheries routinely use driftnets exceeding the EU size limit of 2.5 km/boat. This illegal drift-netting is a major issue, especially in important feeding and breeding grounds for cetaceans.

However, the tuna industry has achieved successes in reversing cetacean bycatch trends. International recognition of the problem of cetacean bycatch in tuna fishing led to the Agreement on the International Dolphin Conservation Program in 1999 and overall there has been a dramatic reduction in death rates. In particular, dolphin bycatch in tuna fishing in the East Tropical Pacific has dropped from 500,000 per year in 1970 to 100,000 per year in 1990 to 3,000 per year in 1999 to 1,000 per year in 2006.

Cetaceans at Risk

Bycatch is recognized as a primary threat to all cetaceans. The following cetaceans are at high risk for entanglement in gillnets:

Atlantic Humpback Dolphins

The Atlantic humpback dolphin (*Sousa teuszii*) is endemic to West Africa. Several stocks have been identified with numbers ranging from tens to a few hundred. Abundance estimates are lacking. Gaps in the species range and hence distribution is evident. Bycatch is only documented in a few West African countries. Surveys and evaluations need to be conducted to determine the presence/ absence of humpback dolphins in their historical range. Conservation measures need to be implemented to save this species. Because many people live off the sea, it is not feasible to have complete gillnet closures. Some areas may be designated as off-limits to gillnet fisheries. Eco-tourism may be implemented successfully because of high species diversity.

Baleen Whales

Baleen whales, *Mysticeti*, are often taken in gill-nets and in fisheries that use vertical lines to mark traps and pots. Large cetaceans such as humpback and right whales may carry off gear after entanglement. This explains the large scars borne by whales along the U.S. Atlantic coast. Analyses show that 50-70% of Gulf of Maine humpback whales, *Megaptera novaeangliae*, and North Atlantic Right Whale, *Eubalaena glacialis*, have been entangled at least once in their lifetime. The North Atlantic right whale is one of the most endangered large cetaceans and only 300-350 individuals remain. Minke Whales, *Balaenoptera acutorostrata*, are also at risk.

North Atlantic Right Whale mother and calf.

Burmeister's Porpoises

The Burmeister's porpoise (*Phocoena spinipinnis*) is one of three cetaceans that are most often bycaught in Peru and Chile. Several thousand porpoises are caught each year in Peru alone. Bycatch is a frequent occurrence for this species because of the inability to detect them in the water. Surveys have shown that bycatch remains a concern in that area today and it is unknown whether or not the population is declining. Data, conservation measures and awareness are lacking. These porpoises are cryptic making surveying a challenge . It is also difficult to estimate bycatch because the sale of porpoise meat is no longer available at markets.

Commerson's Dolphins

A Commerson's Dolphin in an aquarium.

The expanding trawl fisheries devastated the Commerson's dolphin (*Cephalorhynchus commersonii*) populations in Patagonia. Trawl fisheries greatly expanded for twenty years until they crashed in 1997. Pelagic squid fisheries took over which use pelagic trawls that are harmful to dusky, short-beaked common dolphins, and Commerson's dolphins. There are approximately 21,000 Commerson's dolphins remaining today. Two stocks have been identified in the population but genetic information and bycatch levels are unknown. With anchovy fisheries expanding, it is imperative to assess the Commerson's dolphin population before these fisheries grow. The seasonal operation of in-shore gillnet fisheries are known to involve bycatch of cetaceans. Presently, there are no known estimates of gillnet bycatch. The bycatch problem in Argentina is political in nature. Improvements in fishing technology, awareness, and a large scale survey of Commerson's dolphin populations and the impact of bycatch is essential.

La Plata Dolphins

The La Plata or Franciscana dolphin (*Pontoporia blainvillei*) is the most threatened small cetacean in the southwest Atlantic Ocean due to bycatch. They are only found in the coastal waters of Argentina, Brazil, and Uruguay. This species has been divided into four ranges (FMU's: Franciscana Management Units) for management and conservation purposes. These populations are genetically different. Mortality rates are 1.6% for FMU 4 and 3.3% for FMU 3 but it is unknown whether these estimates are accurate. Aerial surveys have proven inconclusive so far as to the population numbers of franciscanas. To rectify this situation, more surveys are needed as well as political commitment, awareness campaigns and bycatch mitigation techniques.

Harbour Porpoises

There is substantial incidental catches in fishing operations. Often, the Harbour Porpoise (*Phocoena phocoena*) is killed by incidental by-catch (10, 11, 12). Gillnets pose a serious threat to the harbour porpoise as they are extremely susceptible to entanglement. A study by Caswell et al. in the western North Atlantic combined the mean annual rate of increase of the harbour porpoise with the uncertainty of incidental mortality and population size. It was found that the incidental mortality exceeds critical values and therefore by-catch is a significant threat to the harbour porpoise. Harbour porpoises become entangled in nets due to their inability to detect the nets before collision. In 2001, 80 harbour porpoises were killed in salmon gillnet fisheries in British Columbia, Canada.

Hector's and Maui's Dolphins

Hector's dolphins have a unique rounded dorsal fin.

In New Zealand, these dolphins have a high rate of entanglement. Hector's dolphin (*Cephalorhynchus hectori*) is endemic to the coastal waters New Zealand and there are about 7,400 in abundance. A small population of Hector's dolphins is isolated on the west coast of the island and have been declared a subspecies called Maui's Dolphin. Maui's dolphins (*Cephalofhynchus hectori maui*) are often caught in set nets and pair trawlers resulting in less than 100 left in the wild. For protection, a section of the dolphin's range on the west coast has been closed to gillnet fisheries.

Indo-Pacific Humpback and Bottlenose Dolphins

Drift and bottom-set gillnets are the biggest conservation threat to these dolphins in the Indian Ocean. There have only been assessments in some areas, such as Zanzibar. Hunting, until 1996, reduced the population and contributed to its decline. Now hunting has been replaced with eco-tourism. It was estimated in 2001 that there are 161 bottlenose dolphins (*Tursiops aduncus*) and 71 Indo-Pacific Humpback Dolphin (*Sousa chinensis*) that are left based on photo-identification mark-recapture techniques. A study on bycatch revealed over 160 incidences of bycatch since 2000. Approximately 30% of bycatch is in drift and bottom-set gillnets. Mortality is about 8% and 5.6% for bottlenose and humpback dolphins respectively . The mitigation of bycatch is imperative for these species and eco-tourism.

Irrawaddy Dolphins

Based on a survey in 2001, fewer than 70 Irrawaddy dolphins (*Orcaella brevirostris*) left in the upper region of the Malampaya Sound in the Philippines and 69 individuals in the Mekong River. They have been severely impacted by lift nets, and crab gear and they are critically endangered. It is estimated that mortality from bycatch may be greater than 4.5% in Malampaya Sound and 5.8% in the Mekong River. The population is declining dramatically. Current bycatch levels are unsustainable and bycatch reduction measures as well as long-term systematic monitoring are urgently required. The elimination of gillnets from areas of high use is needed and economic incentives need to be provided to the local people.

Spinner and Fraser's Dolphins

Spinner dolphins.

In the Philippines, tuna driftnet fisheries have a substantial impact on the populations. One tuna fishery alone kills 400 Spinner Dolphin (*Stenella longirostris*) and Fraser's dolphins (*Lagenodelphis hosei*) each year. Round-haul nets are an even greater concern with a bycatch of up to 3000 dolphins per year. Dolphins that are bycaught often end up as shark bait for longline fisheries.

There is not enough data to conclude total bycatch for the Philippines. Initial assessment indicates that bycatch is not sustainable. Monitoring of dolphin populations and fisheries is urgently needed.

Yangtze River Dolphins and Finless Porpoises

Illustration of a Baiji dolphin.

The Yangtze River or Baiji dolphin (*Lipotes vexillifer*) is the most endangered cetacean and is only found in the Yangtze River, China. A survey conducted in 1997 found only thirteen dolphins. The Yangtze River finless porpoise (*Neophocaena phocaenoides asiaeorientalis*) also lives in the Yangtze River. Abundance has declined and there are fewer than 2000 dolphins left. This may be due, in part, to the construction of the Three Gorges Dam which covers a significant amount of the dolphin's habitat. Both species are often subject to entanglement in gillnets.

Vaquita

The vaquita (*Phocoena sinus*) is highly endangered and is endemic to the upper Gulf of California, Mexico. They are killed in both gillnets and trawl nets from commercial and artisanal fishing. There are presently less than 600 vaquitas left in the Gulf of California.

Mitigating Bycatch

Acoustic Deterrent Devices

The use of acoustic alarms to mitigate by-catch and also to protect aquaculture sites has been proposed but has advantages and risks associated with the alarms. Acoustic deterrent devices, or pingers, have reduced the number of cetaceans caught in gill nets. Harbour porpoises have been effectively excluded from bottom-set gill nets during many experiments for instance in the Gulf of Maine, along the Olympic Peninsula, in the Bay of Fundy, and in the North Sea. All of these studies show up to a 90% decrease in harbour porpoise bycatch. Pingers work because they produce a sound that is aversive (20; 15). There has been a recent re-evaluation of the potential of pingers and their use in other fisheries due to their growing success. An experiment on the California drift gill net fishery demonstrated how acoustic pingers reduce marine mammal bycatch. It was shown that bycatch was significantly reduced for common dolphins and sea lions. Bycatch rates were also lower for other cetacean species like the Northern right whale dolphin, Pacific white-sided dolphin, Risso's dolphin and Dall's Porpoise. It is agreed upon that the more pingers on a net, the less bycatch. There was a 12-fold decrease in common dolphin entanglement using a net with 40 pingers. However, the widespread use of pingers along coastlines effectively excludes cetaceans such as porpoises from prime habitat and resources. Cetaceans which are extremely sensitive to noise are effectively being driven from their preferred coastal habitats by the use of acoustic devices. In poorer quality habitat, harbour porpoises are subjected to increased competition for resources. This situation is recognized as range contraction which can be a result of climate change, anthropogenic activity, or population decline. Large scale range contractions are considered indicative of impending extinction. A similar form of deterrent is noise pollution originating from vessel traffic.

Barium Sulfate

A promising gillnet that is effective in reducing bycatch for harbor porpoises contains barium sulfate. These nets are detected at a greater distance than conventional nets because the barium sulfate reflects the echolocation signal, and also renders the nets more visible. Barium sulfate makes the nets stiffer if it is added at high concentration. All three factors: echo reflectivity, stiffness, and visibility may be important in reducing bycatch. Fish takes in the Bay of Fundy were normal, except for haddock takes, which were down by 3-5%. The advantage of this approach is that it is passive and thus does not require batteries, and there is no "dinner bell" effect. The potential advantage of these nets is greatest in the artisanal fishery. NOAA would like further testing to verify the effectiveness of the nets.

Fishing Regulations and Management

Management and regulation are lacking in many fisheries today. Management measures are urgently needed to monitor fisheries (and illegal fisheries) to protect cetaceans. Efforts to document bycatch should focus on gill-net fisheries because cetaceans are more likely to be caught in gill-nets. Conservation efforts should be directed to areas where marine mammal bycatch is high but where no infrastructure exists to assess the impact. There is a lack of reporting on a global scale of cetacean bycatch.

In the U.S. the Marine Mammal Protection Act prohibits the use and sale of marine mammals captured by fisheries. Similar legislation prohibits the use and sale of marine mammals in other countries. A marine mammal mortality monitoring program for commercial fisheries occurs in the U.S. where "Take Reduction Teams" observe the extent of bycatch and then formulate strategies to reduce bycatch and Take Reduction Plans are put into place.

Temporary Closure

Temporary closure of fisheries during the short period of the year when cetaceans are migrating through the area would decrease bycatch significantly.

Observers on Boat

Observers on fishing vessels to spot cetaceans in the water so that they can be avoided.

In the U.S.

Some programs like Earth Island Institute's Dolphin Safe Label certification claim to require certification from onboard observers. However, the only fishery in the world where independent scientific observers certify whether or not a dolphin has been harmed is the Eastern Tropical Pacific, home to the AIDCP Treaty program. For all other tuna fisheries of the world, the efficacy of onboard observer certification has come under increasing scrutiny as such programs have proven indefensible or unmanageable:

In an interview with Radio Australia last year, Mark Palmer of EII confirmed that it is mostly the case that EII monitors do not go on board of the vessels, and their organization does not have the kind of resources to put observers on the "many thousands" of ships that are out there catching tuna.

Additionally, environmental groups have criticized Earth Island Institute's support of U.S. policies that do not require independent, on-board observation and instead only rely on self-certification by fishing captains, and that even where they may at some point in the future require independent observers, the lack of uniformity in tracing and verifying certifications in different countries means non-certified products can become certified if they are simply taken to the right port.

Other Ways of Mitigating Bycatch

- Implement gear technology (changes in fishing gear and practices) documented to mitigate cetacean bycatch

- Buy tuna and other seafood that has a dolphin safe label.

- Buy Sustainable seafood. To find out which seafood is produced sustainably (i.e. using cetacean friendly gear), refer to World Wildlife Fund Global to access worldwide sustainable seafood guides

- Support sustainable seafood companies and restaurants

- Raise international awareness to assess, monitor and mitigate bycatch problems

- Create legislation on responsible fishing practices.

- Develop and promote industry adoption of "Best Practice Guidelines" for fishing operations

Aquatic Locomotion

Aquatic locomotion is biologically propelled motion through a liquid medium. The simplest propulsive systems are composed of cilia and flagella. Swimming has evolved a number of times in a range of organisms including arthropods, fish, molluscs, reptiles, birds, and mammals.

Evolution of Swimming

Swimming evolved a number of times in unrelated lineages, and the evolutionary pressures leading to its adoption are unknown. Supposed jellyfish fossils occur in the Ediacaran, but the first free-swimming animals appear in the Early to Middle Cambrian. These are mostly related to the arthropods, and include the Anomalocaridids, which swam by means of lateral lobes in a fashion reminiscent of today's cuttlefish. Cephalopods joined the ranks of the nekton in the late Cambrian, and chordates were probably swimming from the Early Cambrian. Many terrestrial animals retain some capacity to swim, however some have returned to the water and developed the capacities for aquatic locomotion.

Micro-organisms

Ciliates

Ciliates use small flagella called cilia to move through the water. One ciliate will generally have hundreds to thousands of cilia that are densely packed together in arrays. During movement, an individual cilium deforms using a high-friction power stroke followed by a low-friction recovery stroke. Since there are multiple cilia packed together on an individual organism, they display collective behavior in a metachronal rhythm. This means the deformation of one cilium is in phase with the deformation of its neighbor, causing deformation waves that propagate along the surface of the organism. These propagating waves of cilia are what allow the organism to use the cilia in a coordinated manner to move. A typical example of a ciliated microorganism is the *Paramecium*, a one-celled, ciliated protozoan covered by thousands of cilia. The cilia beating together allow the *Paramecium* to propel through the water at speeds of 500 micrometers per second.

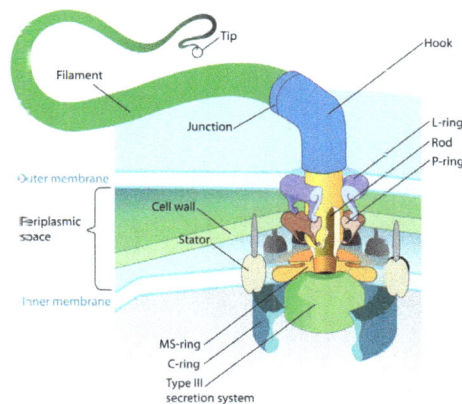

The flagellum of a Gram-negative bacteria is rotated by a molecular motor at its base

Salmon spermatozoa for artificial propagation

Flagellates

Certain organisms such as bacteria and animal sperm have flagellum which have developed a way to move in liquid environments. A rotary motor model shows that bacteria uses the protons of an electrochemical gradient in order to move their flagella. Torque in the flagella of bacteria is created by particles that conduct protons around the base of the flagellum. The direction of rotation of the flagella in bacteria comes from the occupancy of the proton channels along the perimeter of the flagellar motor.

Movement of sperm is called sperm motility. The middle of the mammalian spermatozoon contains mitochondria that power the movement of the flagellum of the sperm. The motor around the base produces torque, just like in bacteria for movement through the aqueous environment.

Pseudopodia

Movement using a pseudopod is accomplished through increases in pressure at one point on the cell membrane. This pressure increase is the result of actin polymerization between the cortex and the membrane. As the pressure increases the cell membrane is pushed outward creating the pseudopod. When the pseudopod moves outward, the rest of the body is pulled forward by cortical tension. The result is cell movement through the fluid medium. Furthermore, the direction of movement is determined by chemotaxis. When chemoattraction occurs in a particular area of the cell membrane, actin polymerization can begin and move the cell in that direction. An excellent example of an organism that utilizes pseudopods is *Naegleria fowleri*.

Invertebrates

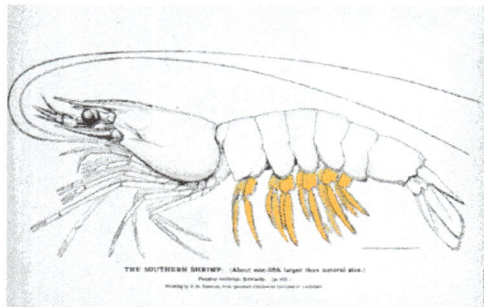

Shrimp paddle with special swimming legs (pleopods)

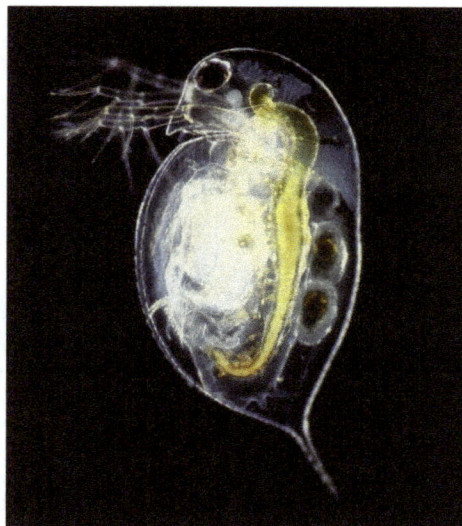

Daphnia swims by beating its antennae

Among the radiata, jellyfish and their kin, the main form of swimming is to flex their cup shaped bodies. All jellyfish are free-swimming, although many of these spend most of their time swimming passively. Passive swimming is akin to gliding; the organism floats, using currents where it can, and does not exert any energy into controlling its position or motion. Active swimming, in contrast, involves the expenditure of energy to travel to a desired location.

In bilateria, there are many methods of swimming. The arrow worms (chaetognatha) undulate their finned bodies, not unlike fish. Nematodes swim by undulating their fin-less bodies. Some Arthropod groups can swim - including many crustaceans. Most crustaceans, such as shrimp, will usually swim by paddling with special swimming legs (pleopods). Swimming crabs swim with modified walking legs (pereiopods). Daphnia, a crustacean, swims by beating its antennae instead.

There are also a number of forms of swimming molluscs. Many free-swimming sea slugs, such as sea angels, flap fin-like structures. Some shelled molluscs, such as scallops can briefly swim by clapping their two shells open and closed. The molluscs most evolved for swimming are the cephalopods.

Among the Deuterostomia, there are a number of swimmers as well. Feather stars can swim by undulating their many arms . Salps move by pumping waters through their gelatinous bodies. The deuterosomes most evolved for swimming are found among the vertebrates, notably the fish.

Jet Propulsion

Jet propulsion is a method of aquatic locomotion where animals fill a muscular cavity and squirt out water to propel them in the opposite direction of the squirting water. Most organisms are equipped with one of two designs for jet propulsion; they can draw water from the rear and expel it from the rear, such as jellyfish, or draw water from front and expel it from the rear, such as salps. Filling up the cavity causes an increase in both the mass and drag of the animal. Because of the expanse of the contracting cavity, the animal's velocity fluctuates as it moves through the water, accelerating while expelling water and decelerating while vacuuming water. Even though these fluctuations in drag and mass can be ignored if the frequency of the jet-propulsion cycles is high enough, jet-propulsion is a relatively inefficient method of aquatic locomotion.

Octopuses swim headfirst, with arms trailing behind

Jellyfish pulsate their bell for a type of jet locomotion

Scallops swim by clapping their two shells open and closed

All cephalopods can move by jet propulsion, but this is a very energy-consuming way to travel compared to the tail propulsion used by fish. The relative efficiency of jet propulsion decreases further as animal size increases. Since the Paleozoic, as competition with fish produced an environment where efficient motion was crucial to survival, jet propulsion has taken a back role, with fins and tentacles used to maintain a steady velocity. The stop-start motion provided by the jets, however, continues to be useful for providing bursts of high speed - not least when capturing prey or avoiding predators. Indeed, it makes cephalopods the fastest marine invertebrates,[Preface] and they can outaccelerate most fish. Oxygenated water is taken into the mantle cavity to the gills and through muscular contraction of this cavity, the spent water is expelled through the hyponome, created by a fold in the mantle. Motion of the cephalopods is usually backward as water is forced out anteriorly through the hyponome, but direction can be controlled somewhat by pointing it in different directions. Most cephalopods float (i.e. are neutrally buoyant), so do not need to swim to remain afloat. Squid swim more slowly than fish, but use more power to generate their speed. The loss in efficiency is due to the amount of water the squid can accelerate out of its mantle cavity.

Jellyfish use a one-way water cavity design which generates a phase of continuous cycles of jet-propulsion followed by a rest phase. The Froude efficiency is about 0.09, which indicates a very costly method of locomotion. The metabolic cost of transport for jellyfish is high when compared to a fish of equal mass.

Other jet-propelled animals have similar problems in efficiency. Scallops, which use a similar design to jellyfish, swim by quickly opening and closing their shells, which draws in water and expels it from all sides. This locomotion is used as a means to escape predators such as starfish. Afterwards, the shell acts as a hydrofoil to counteract the scallop's tendency to sink. The Froude efficiency is low for this type of movement, about 0.3, which is why it's used as an emergency escape mechanism from predators. However, the amount of work the scallop has to do is mitigated by the elastic hinge that connects the two shells of the bivalve. Squids swim by drawing water into their mantle cavity and expelling it through their siphon. The Froude efficiency of their jet-propulsion system is around 0.29, which is much lower than a fish of the same mass.

Much of the work done by scallop muscles to close its shell is stored as elastic energy in abductin tissue, which acts as a spring to open the shell. The elasticity causes the work done against the water to be low because of the large openings the water has to enter and the small openings the water has to leave. The inertial work of scallop jet-propulsion is also low. Because of the low inertial work, the energy savings created by the elastic tissue is so small that it's negligible. Medusae can also use their elastic mesoglea to enlarge their bell. Their mantle contains a layer of muscle sandwiched between elastic fibers. The muscle fibers run around the bell circumferentially while the elastic fibers run through the muscle and along the sides of the bell to prevent lengthening. After making a single contraction, the bell vibrates passively at the resonant frequency to refill the bell. However, in contrast with scallops, the inertial work is similar to the hydrodynamic work due to how medusas expel water - through a large opening at low velocity. Because of this, the negative pressure created by the vibrating cavity is lower than the positive pressure of the jet, meaning that inertial work of the mantle is small. Thus, jet-propulsion is shown as an inefficient swimming technique.

Fish

Open water fish, like this Atlantic bluefin tuna, are usually streamlined for straightline speed, with a deeply forked tail and a smooth body shaped like a spindle tapered at both ends.

Many reef fish, like this queen angelfish, have a body flattened like a pancake, with pectoral and pelvic fins that act with the flattened body to maximize manoeuvrability.

Many fish swim through water by creating undulations with their bodies or oscillating their fins. The undulations create components of forward thrust complemented by a rearward force, side forces which are wasted portions of energy, and a normal force that is between the forward thrust and side force. Different fish swim by undulating different parts of their bodies. Eel-shaped fish undulate their entire body in rhythmic sequences. Streamlined fish, such as salmon, undulate the caudal portions of their bodies. Some fish, such as sharks, use stiff, strong fins to create dynamic lift and propel themselves. It is common for fish to use more than one form of propulsion, although they will display one dominant mode of swimming Gait changes have even been observed in juvenile reef fish of various sizes. Depending on their needs, fish can rapidly alternate between synchronized fin beats and alternating fin beats

According to *Guinness World Records 2009*, *H. zosterae* (the dwarf seahorse) is the slowest moving fish, with a top speed of about 5 feet (150 cm) per hour. They swim very poorly, rapidly fluttering a dorsal fin and using pectoral fins (located behind their eyes) to steer. Seahorses have no caudal fin.

Body-caudal Fin (BCF) Propulsion

- Anguilliform: Anguilliform swimmers are typically slow swimmers. They undulate the majority of their body and use their head as the fulcrum for the load they are moving. At any point during their undulation, their body has an amplitude between 0.5-1.0 wavelengths. The amplitude that they move their body through allows them to swim backwards. Anguilliform locomotion is usually seen in fish with long, slender bodies like eels, lampreys, oarfish, and a number of catfish species.

- Subcarangiform, Carangiform, Thunniform: These swimmers undulate the posterior half of their body and are much faster than anguilliform swimmers. At any point while they are swimming, a wavelength <1 can be seen in the undulation pattern of the body. Some Carangiform swimmers include nurse sharks, bamboo sharks, and reef sharks. Thunniform swimmers are very fast and some common Thunniform swimmers include tuna, white sharks, salmon, jacks, and mako sharks. Thunniform swimmers only undulate their high aspect ratio caudal fin, so they are usually very stiff to push more water out of the way.

- Ostraciiform: Ostraciiform swimmers oscillate their caudal region, making them relatively slow swimmers. Boxfish, torpedo rays, and momyrs employ Ostraciiform locomotion. The cow fish uses Osctraciiform locomotion to hover in the water column.

Median Paired Fin (MPF) Propulsion

- Tetraodoniform, Balistiform, Diodontiform: These swimmers oscillate their median fins. They are typically slow swimmers, and some notable examples include the oceanic sunfish (which has extremely modified anal and dorsal fins), puffer fish, and triggerfish.

- Rajiform, Amiiform, Gymnotiform: This locomotory mode is accomplished by undulation of the pectoral and median fins. During their undulation pattern, a wavelength >1 can be seen in their fins. They are typically slow to moderate swimmers, and some examples include rays, bowfin, and knife fishes. The black ghost knife fish is a Gymnotiform swimmer that has a very long ventral ribbon fin. Thrust is produced by passing waves down the ribbon fin while the body remains rigid. This also allows the ghost knife fish to swim in reverse.

- Labriform: Labriform swimmers are also slow swimmers. They oscillate their pectoral fins to create thrust. Oscillating fins create thrust when a starting vortex is shed from the trailing edge of the fin. As the foil departs from the starting vortex, the effect of that vortex diminishes, while the bound circulation remains, producing lift. Labriform swimming can be viewed as continuously starting and stopping. Wrasses and surf perch are common Labriform swimmers

Hydrofoils

Hydrofoils, or fins, are used to push against the water to create a normal force to provide thrust, propelling the animal through water. Sea turtles and penguins beat their paired hydrofoils to create lift. Some paired fins, such as pectoral fins on leopard sharks, can be angled at varying degrees to allow the animal to rise, fall, or maintain its level in the water column. The reduction of fin surface area helps to minimize drag, and therefore increase efficiency. Regardless of size of the animal, at any particular speed, maximum possible lift is proportional to (wing area) x (speed)^2. Dolphins and whales have large, horizontal caudal hydrofoils, while many fish and sharks have vertical caudal hydrofoils. Porpoising (seen in cetaceans, penguins, and pinnipeds) may save energy if they are moving fast. Since drag increases with speed, the work required to swim unit distance is greater at higher speeds, but the work needed to jump unit distance is independent of speed. Seals propel themselves through the water with their caudal tail, while sea lions create thrust solely with their pectoral flippers.

The leopard shark angles its pectoral fins so they behave as hydrofoils to control the animal's pitch

Drag Powered Swimming

The slowest-moving fishes are the sea horses, often found in reefs

As with moving through any fluid, friction is created when molecules of the fluid collide with organism. The collision causes drag against moving fish, which is why many fish are streamlined in shape. Streamlined shapes work to reduce drag by orienting elongated objects parallel to the force of drag, therefore allowing the current to pass over and taper off the end of the fish. This streamlined shape allows for more efficient use of energy locomotion. Some flat-shaped fish can take advantage of pressure drag by having a flat bottom surface and curved top surface. The pressure drag created allows for the upward lift of the fish.

Appendages of aquatic organisms propel them in two main and biomechanically extreme mechanisms. Some use lift powered swimming, which can be compared to flying as appendages flap like wings, and reduce drag on the surface of the appendage. Others use drag powered swimming, which can be compared to oars rowing a boat, with movement in a horizontal plane, or paddling, with movement in the parasagittal plane.

Drag swimmers use a cyclic motion where they push water back in a power stroke, and return their limb forward in the return or recovery stroke. When they push water directly backwards, this moves their body forward, but as they return their limbs to the starting position, they push water forward, which will thus pull them back to some degree, and so opposes the direction that the body is heading. This opposing force is called drag. The return-stroke drag causes drag swimmers to employ different strategies than lift swimmers. Reducing drag on the return stroke is essential for optimizing efficiency. For example, ducks paddle through the water spreading the webs of their feet as they move water back, and then when they return their feet to the front they pull their webs together to reduce the subsequent pull of water forward. The legs of water beetles have little hairs which spread out to catch up and move water back in the power stroke, but lay flat as the appendage moves forward in the return stroke. Also, the water beetle's legs have a side that is wider and is held perpendicular to the motion when pushing backward, but the leg is then rotated when the limb is to return forward, so that the thinner side will catch up less water.

Drag swimmers experience a lessened efficiency in swimming due to resistance which affects their optimum speed. The less drag a fish experiences, the more it will be able to maintain higher speeds. Morphology of the fish can be designed to reduce drag, such as streamlining the body. The cost of transport is much higher for the drag swimmer, and when deviating from its optimum speed, the drag swimmer is energetically strained much more than the lift swimmer. There are natural processes in place to optimize energy use, and it is thought that adjustments of metabolic rates can compensate in part for mechanical disadvantages.

Semi-aquatic animals compared to fully aquatic animals exhibit exacerbation of drag. Design that allows them to function out of the water limits the efficiency possible to be reached when in the water. In water swimming at the surface exposes them to resistive wave drag and is associated with a higher cost than submerged swimming. Swimming below the surface exposes them to resistance due to return strokes and pressure, but primarily friction. Frictional drag is due to fluid viscosity and morphology characteristics. Pressure drag is due to the difference of water flow around the body and is also affected by body morphology. Semi-aquatic organisms encounter increased resistive forces when in or out of the water, as they are not specialized for either habitat. The morphology of otters and beavers, for example, must meet needs for both environments. Their fur decreases streamlining and creates additional drag. The platypus may be a good example of an intermediate between drag and lift swimmers because it has been shown to have a rowing mechanism which is similar to lift-based pectoral oscillation. The limbs of semi-aquatic organisms are reserved for use on land and using them in water not only increases the cost of locomotion, but limits them to drag-based modes. Although they are less efficient, drag swimmers are able to produce more thrust at low speeds than lift swimmers. They are also thought to be better for maneuverability due to the large thrust produced.

Amphibians

Most of the Amphibia have a larval state, which has inherited anguilliform motion, and a laterally compressed tail to go with it, from fish ancestors. The corresponding tetrapod adult forms, even in the tail-retaining sub-class Urodeles, are sometimes aquatic to only a negligible extent (as in the genus Salamandra, whose tail has lost its suitability for aquatic propulsion), but the majority of Urodeles, from the newts to the giant salamander Megalobatrachus, retain a laterally compressed tail for a life that is aquatic to a considerable degree, which can use in a carangiform motion.

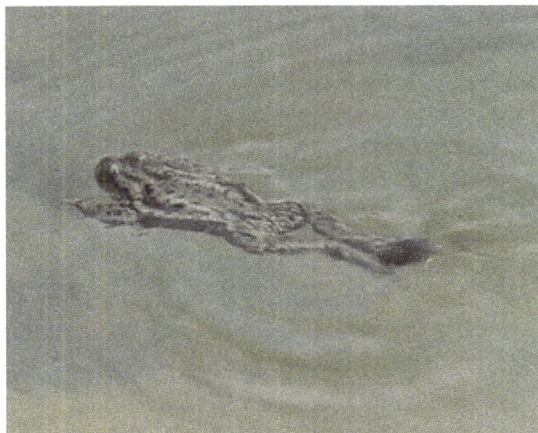

Common toad (*Bufo bufo*) swimming

Of the tailless amphibians (the frogs and toads of the sub-class Anura) the majority are aquatic to an insignificant extent in adult life, but in that considerable minority that are mainly aquatic we encounter for the first time the problem of adapting the tailless-tetrapod structure for aquatic propulsion. The mode that they use is unrelated to any used by fish. With their flexible back legs and webbed feet they execute something close to the leg movements of a human 'breast stroke,' rather more efficiently because the legs are better streamlined.

Reptiles

Nile crocodile (*Crocodylus niloticus*) swimming

From the point of view of aquatic propulsion, the descent of modern members of the class Reptilia from archaic tailed Amphibia is most obvious in the case of the order Crocodilia (crocodiles and alligators), which use their deep, laterally compressed tails in an essentially carangiform mode of propulsion.

Terrestrial snakes, in spite of their 'bad' hydromechanical shape with roughly circular cross-section and gradual posterior taper, swim fairly readily when required, by an anguilliform propulsion.

Cheloniidae (true turtles) have found a beautiful solution to the problem of tetrapod swimming through the development of their forelimbs into flippers of high-aspect-ratio wing shape, with which they imitate a bird's propulsive mode more accurately than do the eagle-rays themselves.

Immature Hawaiian green sea turtle in shallow waters

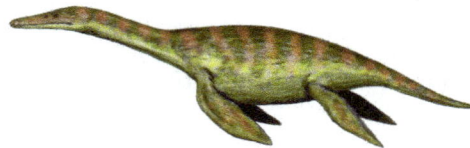

Macroplata

Fin and Flipper Locomotion

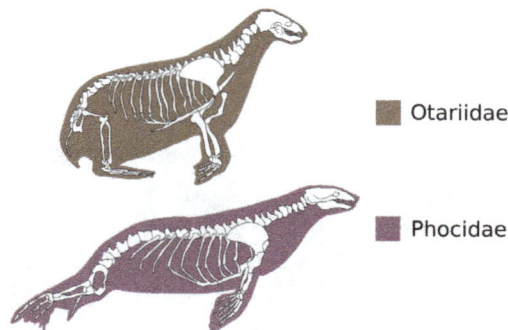

Otariidae

Phocidae

Comparative skeletal anatomy of a typical otariid seal and a typical phocid seal

Aquatic reptiles such as sea turtles and extinct species like Pliosauroids predominantly use their pectoral flippers to propel themselves through the water and their pelvic flippers for maneuvering. During swimming they move their pectoral flippers in a dorso-ventral motion, causing forward motion. During swimming, they rotate their front flippers to decrease drag through the water column and increase efficiency. Newly hatched sea turtles exhibit several behavioral skills that help orient themselves towards the ocean as well as identifying the transition from sand to water. If rotated in the pitch, yaw or roll direction, the hatchlings are capable of counteracting the forces acting upon them by correcting with either their pectoral or pelvic flippers and redirecting themselves towards the open ocean.

Among mammals otariids (fur seals) swim primarily with their front flippers, using the rear flippers for steering, and phocids (true seals) move the rear flippers laterally, pushing the animal through the water.

Escape Reactions

Some arthropods, such as lobsters and shrimps, can propel themselves backwards quickly by flicking their tail, known as lobstering or the caridoid escape reaction.

Varieties of fish, such as teleosts, also use fast-starts to escape from predators. Fast-starts are characterized by the muscle contraction on one side of the fish twisting the fish into a C-shape. Afterwards, muscle contraction occurs on the opposite side to allow the fish to enter into a steady swimming state with waves of undulation traveling alongside the body. The power of the bending motion comes from fast-twitch muscle fibers located in the central region of the fish. The signal to perform this contraction comes from a set of Mauthner cells which simultaneously send a signal to the muscles on one side of the fish. Mauthner cells are activated when something startles the fish and can be activated by visual or sound-based stimuli.

Fast-starts are split up into three stages. Stage one, which is called the preparatory stroke, is characterized by the initial bending to a C-shape with small delay caused by hydrodynamic resistance. Stage two, the propulsive stroke, involves the body bending rapidly to the other side, which may occur multiple times. Stage three, the rest phase, cause the fish to return to normal steady-state swimming and the body undulations begin to cease. Large muscles located closer to the central portion of the fish are stronger and generate more force than the muscles in the tail. This asymmetry in muscle composition causes body undulations that occur in Stage 3. Once the fast-start is completed, the position of the fish has been shown to have a certain level of unpredictability, which helps fish survive against predators.

The rate at which the body can bend is limited by resistance contained in the inertia of each body part. However, this inertia assists the fish in creating propulsion as a result of the momentum created against the water. The forward propulsion created from C-starts, and steady-state swimming in general, is a result of the body of the fish pushing against the water. Waves of undulation create rearward momentum against the water providing the forward thrust required to push the fish forward.

Efficiency

The Froude propulsion efficiency is defined as the ratio of power output to the power input:

$$nf = 2U_1 / (U_1 + U_2)$$

where U1 = free stream velocity and U2 = jet velocity. A good efficiency for carangiform propulsion is between 50 and 80%.

Minimizing Drag

Pressure differences occur outside the boundary layer of swimming organisms due to disrupted flow around the body. The difference on the up- and down-stream surfaces of the body is pressure drag, which creates a downstream force on the object. Frictional drag, on the other hand, is a result of fluid viscosity in the boundary layer. Higher turbulence causes greater frictional drag.

Reynolds number (Re) is the measure of the relationships between inertial and viscous forces in flow ((animal's length x animal's velocity)/kinematic viscosity of the fluid). Turbulent flow can be found at higher Re values, where the boundary layer separates and creates a wake, and laminar flow can be found at lower Re values, when the boundary layer separation is delayed, reducing wake and kinetic energy loss to opposing water momentum.

The body shape of a swimming organism affects the resulting drag. Long, slender bodies reduce pressure drag by streamlining, while short, round bodies reduce frictional drag; therefore, the optimal shape of an organism depends on its niche. Swimming organisms with a fusiform shape are likely to experience the greatest reduction in both pressure and frictional drag.

Wing shape also affects the amount of drag experienced by an organism, as with different methods of stroke, recovery of the pre-stroke position results in the accumulation of drag.

High-speed ram ventilation creates laminar flow of water from the gills along the body of an organism.

The secretion of mucus along the organism's body surface, or the addition of long-chained polymers to the velocity gradient, can reduce frictional drag experienced by the organism.

Buoyancy

Many aquatic/marine organisms have developed organs to compensate for their weight and control their buoyancy in the water. These structures, make the density of their bodies very close to that of the surrounding water. Some hydrozoans, such as siphonophores, has gas-filled floats; the Nautilus, Sepia, and Spirula (Cephalopods) have chambers of gas within their shells; and most teleost fish and many lantern fish (Myctophidae) are equipped with swim bladders. Many aquatic and marine organisms may also be composed of low-density materials. Deep-water teleosts, which do not have a swim bladder, have few lipids and proteins, deeply ossified bones, and watery tissues that maintain their buoyancy. Some sharks' livers are composed of low-density lipids, such as hydrocarbon squalene or wax esters (also found in Myctophidae without swim bladders), which provide buoyancy.

Swimming animals that are denser than water must generate lift or adapt a benthic lifestyle. Movement of the fish to generate hydrodynamic lift is necessary to prevent sinking. Often, their bodies act as hydrofoils, a task that is more effective in flat-bodied fish. At a small tilt angle, the lift is greater for flat fish than it is for fish with narrow bodies. Narrow-bodied fish use their fins as hydrofoils while their bodies remain horizontal. In sharks, the heterocercal tail shape drives water downward, creating a counteracting upward force while thrusting the shark forward. The lift generated is assisted by the pectoral fins and upward-angle body positioning. It is supposed that tunas primarily use their pectoral fins for lift.

Buoyancy maintenance is metabolically expensive. Growing and sustaining a buoyancy organ, adjusting the composition of biological makeup, and exerting physical strain to stay in motion demands large amounts of energy. It is proposed that lift may be physically generated at a lower energy cost by swimming upward and gliding downward, in a "climb and glide" motion, rather than constant swimming on a plane.

Temperature

Temperature can also greatly affect the ability of aquatic organisms to move through water. This is because temperature not only affects the properties of the water, but also the organisms in the water, as most have an ideal range specific to their body and metabolic needs.

Q_{10} (temperature coefficient), the factor by which a rate increases at a 10 °C increase in temperature, is used to measure how organisms' performance relies on temperature. Most have increased rates as water becomes warmer, but some have limits to this and others find ways to alter such effects, such as by endothermy or earlier recruitment of faster muscle.

For example, *Crocodylus porosus*, or estuarine crocodiles, were found to increase swimming speed from 15 °C to 23 °C and then to have peak swimming speed from 23 °C to 33 °C. However, performance began to decline as temperature rose beyond that point, showing a limit to the range of temperatures at which this species could ideally perform.

Submergence

The more of the animal's body that is submerged while swimming, the less energy it uses. Swimming on the surface requires two to three times more energy than when completely submerged. This is because of the bow wave that is formed at the front when the animal is pushing the surface of the water when swimming, creating extra drag.

Secondary Evolution

Chinstrap penguin leaping over water

Swimming dog

While tetrapods lost many of their natural adaptations to swimming when they evolved onto the land, many have re-evolved the ability to swim or have indeed returned to a completely aquatic lifestyle.

Primarily or exclusively aquatic animals have re-evolved from terrestrial tetrapods multiple times: examples include amphibians such as newts, reptiles such as crocodiles, sea turtles, ichthyosaurs, plesiosaurs and mosasaurs, marine mammals such as whales, seals and otters, and birds such as penguins. Many species of snakes are also aquatic and live their entire lives in the water. Among invertebrates, a number of insect species have adaptations for aquatic life and locomotion. Examples of aquatic insects include dragonfly larvae, water boatmen, and diving beetles. There are also aquatic spiders, although they tend to prefer other modes of locomotion under water than swimming proper.

Even though primarily terrestrial tetrapods have lost many of their adaptations to swimming, the ability to swim has been preserved or re-developed in many of them. It may never have been completely lost.

Examples are: Some breeds of dog swim recreationally. Umbra, a world record-holding dog, can swim 4 miles (6.4 km) in 73 minutes, placing her in the top 25% in human long-distance swimming competitions. Although most cats hate water, adult cats are good swimmers. The fishing cat is one wild species of cat that has evolved special adaptations for an aquatic or semi-aquatic lifestyle – webbed digits. Tigers and some individual jaguars are the only big cats known to go into water readily, though other big cats, including lions, have been observed swimming. A few domestic cat breeds also like swimming, such as the Turkish Van.

Horses, moose, and elk are very powerful swimmers, and can travel long distances in the water. Elephants are also capable of swimming, even in deep waters. Eyewitnesses have confirmed that camels, including dromedary and Bactrian camels, can swim, despite the fact that there is little deep water in their natural habitats.

Both domestic and wild rabbits can swim. Domestic rabbits are sometimes trained to swim as a circus attraction. A wild rabbit famously swam in an apparent attack on U.S. President Jimmy Carter's boat when it was threatened in its natural habitat.

The guinea pig (or cavy) is noted as having an excellent swimming ability. Mice can swim quite well. They do panic when placed in water, but many lab mice are used in the Morris water maze, a test to measure learning. When mice swim, they use their tails like flagella and kick with their legs.

Many snakes are excellent swimmers as well. Large adult anacondas spend the majority of their time in the water, and have difficulty moving on land.

Many monkeys can naturally swim and some, like the proboscis monkey, crab-eating macaque, and rhesus macaque swim regularly.

Human Swimming

Swimming has been known amongst humans since prehistoric times; the earliest record of swimming dates back to Stone Age paintings from around 7,000 years ago. Competitive swimming started in Europe around 1800 and was part of the first modern 1896 Summer Olympics in Athens, though not in a form comparable to the contemporary events. It was not until 1908 that regulations were implemented by the International Swimming Federation to produce competitive swimming.

Marine Mammal Observer

A marine mammal observer (MMO) is a professional in environmental consulting who specializes in whales and dolphins.

In recent years there has been increased concern for the effect of man-made noise pollution in the ocean, particularly upon cetaceans - which are known to be sensitive to sound. As a result, environmental regulations have been introduced in an attempt to minimise negative impacts on marine wildlife. These guidelines have focused on the oil industry's seismic exploration for offshore oil. They center on the practice of delaying or shutting down the use of air-guns if a whale or dolphin is sighted nearby. An MMO will implement these regulations in the field.

When on board the seismic vessel, the MMO's job is two-fold:

- To spot sensitive wildlife species

- To monitor adherence to the guidelines.

Spotting, and identifying, animals involves long hours of visual surveys. Detecting cetaceans with hydrophones is known as Passive Acoustic Monitoring (PAM), and this is an increasingly common technique used in addition to visual surveys. Ensuring adherence to guidelines requires a thorough knowledge of the regulations, understanding of the operations and the ability to communicate effectively with the crew. MMOs usually have a strong background in marine biology and conservation. Increasingly, the oil industry is employing a 'best practice' attitude to environmental commitment and voluntarily taking on MMOs as independent observers in areas where no government regulations exist. In some circumstances guidelines may be open to interpretation or the environmental conditions unique and the MMO will be called upon to advice on a sensible mitigation protocol.

As well as the seismic exploration industry, MMOs may also be required during; oil rig decommissioning, where disused oil platform pilings on the seabed are removed by large amounts of explosives, marine construction projects and; military trials of powerful new active sonar systems. Typically MMO duties are funded by the company surveying.

Noise from human activity in the ocean environment is likely to increase - and become a bigger environmental issue. Discussion of how to minimize the negative effects of noise upon whales, dolphins and other marine-life will no doubt continue between industry, government agencies, military, environmental organizations and academics. It will be the MMO who puts this into practice in the field.

Marine Mammals and Sonar

Active sonar, the transmission equipment used on some ships to assist with navigation, is detrimental to the health and livelihood of some marine animals. Research has recently shown that beaked and blue whales are sensitive to mid-frequency active sonar and move rapidly away from the source of the sonar, a response that disrupts their feeding and can cause mass strandings. Some marine animals, such as whales and dolphins, use echolocation or "biosonar" systems to locate predators and prey. It is conjectured that active sonar transmitters could confuse these animals and interfere with basic biological functions such as feeding and mating. Study has shown whales experience decompression sickness, a disease that forces nitrogen into gas bubbles in the tissues and is caused by rapid and prolonged surfacing. Although whales were originally thought to be immune to this disease, sonar has been implicated in causing behavioral changes that can lead to decompression sickness.

History

The SOFAR channel (short for sound fixing and ranging channel), or deep sound channel (DSC), is a horizontal layer of water in the ocean centered around the depth at which the speed of sound is at a minimum. The SOFAR channel acts as a waveguide for sound, and low frequency sound waves within the channel may travel thousands of miles before dissipating. This phenomenon is an important factor in submarine warfare. The deep sound channel was discovered and described independently by Dr. Maurice Ewing, and Leonid Brekhovskikh in the 1940s.

Despite the use of the SOFAR channel in naval applications, the idea that animals might make use of this channel was not proposed until 1971. Roger Payne and Douglas Webb calculated that before ship traffic noise permeated the oceans, tones emitted by fin whales could have traveled as far as four thousand miles and still be heard against the normal background noise of the sea. Payne and Webb further determined that, on a quiet day in the pre–ship-propeller oceans, fin whale tones would only have fallen to the level of background noise after traveling thirteen thousand miles, that is, more than the diameter of the Earth.

Early Confusion between Fin Whales and Military Sonar

Before extensive research on whale vocalizations was completed, the low-frequency pulses emitted by some species of whales were often not correctly attributed to them. Dr Payne wrote: "Before it was shown that fin whales were the cause [of powerful sounds], no one could take seriously the idea that such regular, loud, low, and relatively pure frequency tones were coming from within the ocean, let alone from whales." This unknown sound was popularly known by navy acousticians as the *Jezebel Monster*. (*Jezebel* was narrow-band passive long-range sonar.) Some researchers[who?] believed that these sounds could be attributed to geophysical vibrations or an unknown Russian military program, and it wasn't until biologists William Schevill and William A. Watkins proved that whales possessed the biological capacity to emit sounds that the unknown sounds were correctly attributed.

Low Frequency Sonar

The electromagnetic spectrum has rigid definitions for "super low frequency", "extremely low frequency", "low frequency" and "medium frequency". Acoustics does not have a similar standard. The terms "low" and "mid" have roughly-defined historical meanings in sonar, because not many frequencies have been used over the decades. However, as more experimental sonars have been introduced, the terms have become muddled.

American low frequency sonar was originally introduced to the general public in a June 1961 *Time* magazine article, *New A.S.W.* Project Artemis, the low-frequency sonar used at the time, could fill a whole ocean with searching sound and spot anything sizable that was moving in the water. Artemis grew out of a 1951 suggestion by Harvard physicist Frederick V. Hunt (Artemis is the Ancient Greek goddess of the hunt), who convinced Navy anti-submarine experts that submarines could be detected at great distances only by unheard-of volumes of low-pitched sound. At the time, an entire Artemis system was envisioned to form a sort of underwater DEW (*Distant Early Warning*) line to warn the U.S. of hostile submarines. Giant, unattended transducers, powered by cables from land, would be lowered to considerable depths where sound travels best. The *Time* magazine article was published during the maiden voyage of the Soviet submarine *K-19*, which was the first Soviet submarine equipped with ballistic missiles. Four days later the submarine would have the accident that gave it its nickname. The impact on marine mammals by this system was certainly not a consideration. Artemis never became an operational system.

Low-frequency sonar was revived in the early 1980s for military and research applications. The idea that the sound could interfere with whale biologics became widely discussed outside of research circles when Scripps Institute of Oceanography borrowed and modified a military sonar for the Heard Island Feasibility Test conducted in January and February 1991. The sonar modified for the test was an early version of SURTASS deployed in the MV *Cory Chouest*. As a result of this test a "Committee on Low-Frequency Sound and Marine Mammals" was organized by the National Research Council. Their findings were published in 1994, in *Low-Frequency Sound and Marine Mammals: Current Knowledge and Research Needs*.

Long-range transmission does not require high power. All frequencies of sound lose an average of 65dB in the first few seconds before the sound waves strike the ocean bottom. After that the acoustic energy in mid or high-frequency sound is converted into heat, primarily by the epsom salt dissolved in sea water. Very little of low frequency acoustic energy is converted into heat, so the signal can be detected for long ranges. Fewer than five of the transducers from the low frequency active array were used in the Heard Island Feasibility Test, and the sound was detected on the opposite side of the Earth. The transducers were temporarily altered for this test to transmit sound at 50 hertz, which is lower than their normal operating frequency.

A year after the Heard Island Feasibility Test a new low-frequency active sonar was installed in the *Cory Chouest* with 18 transducers instead of 10. An environmental impact statement was prepared for that system.

Mid-frequency Sonar

The term *mid-frequency* sonar is usually used to refer to sonars that project sound in the 3 to 4 kilohertz (kHz) range. Ever since the launch of the USS *Nautilus* (SSN-571) on 17 January 1955 the US Navy knew it was only a matter of time until the other naval powers had their own nuclear submarines. The mid-frequency sonar was developed for anti-submarine warfare against these future boats. The standard post-WWII active sonars (which were usually above 7 kHz) had an insufficient range against this new threat. Active sonar went from a piece of equipment attached to a ship, to a piece of equipment that was central to the design of a ship. They are described in the same 1961 *Time* magazine article by the quote "*the latest shipboard sonar weighs 30 tons and consumes 1,600 times as much power as the standard postwar sonar*". A modern system produced by Lockheed Martin since the early 1980s is the AN/SQQ-89. On June 13, 2001, Lockheed Martin announced that it had delivered its 100th AN/SQQ-89 undersea warfare system to the U.S. Navy.

There was anecdotal evidence that mid-frequency sonar could have adverse effects on whales dating back to the days of whaling. The following story is recounted in a book published in 1995:

Mid Frequency Sonar and Whaling
Source: *Among Whales* by Roger Payne (pg 258) Published 2 June 1995
Another innovation by the whalers was the use of sonar to track whales they were pursuing underwater. But there was a problem; as the boat gained on the whale, the whale started exhaling while still submerged. This produced a cloud of bubbles in the water that reflected sound better than the whale did and made a false target (akin to what a pilot does when releasing metal chaff to create a false radar echo). I suspect that this behavior by whales was simply fortuitous since exhaling while still submerged is simply a means by which a whale can reduce the time it has to remain at the surface, where surface drag will slow it down.   *Whalers quickly discovered that a frequency of **three thousand hertz** seemed to panic the whales, causing them to surface much more often for air, This was a "better" use for sonar because it afforded the whalers more chances to shoot the whales. So they equipped their catcher boats with sonar at that frequency. Of course the sonar also allows the whalers to follow the whale underwater, but that is its secondary use. Its primary use is for scaring whales so that they start "panting" at the surface.*

In 1996 twelve Cuvier's beaked whales beached themselves alive along the coast of Greece while NATO (North Atlantic Treaty Organisation) was testing an active sonar with combined low and mid-range frequency transducers, according to a paper published in the journal *Nature* in 1998. The author established for the first time the link between atypical mass strandings of whales and the use of military sonar by concluding that *although pure coincidence cannot be excluded* there was better than a 99.3% likelihood that sonar testing caused that stranding. He noted that the whales were spread along 38.2 kilometres of coast and were separated by a mean distance of 3.5 km (sd=2.8, n=11). This spread in time and location was atypical, as usually whales mass strand at the same place and at the same time.

At the time that Dr. Frantzis wrote the article he was unaware of several important factors.

- The time correlation was much tighter than he knew. He knew about the test from a notice to mariners which only published that the test would occur over a five-day period within a large area of the ocean. In fact the first time the sonar was turned on was the morning of 12 May 1996, and six whales stranded that afternoon. The next day the sonar was turned on again and another six whales stranded that afternoon. Without knowing the coordinates of the ships he would not have realized that the ship was only about 10–15 miles offshore.

- The sonar being used in the test was an experimental research and development sonar, which was considerably smaller and less powerful than an operational sonar on board a deployed naval vessel. Dr Frantzis believed that wide distribution of the stranded whales indicated that the cause has a large synchronous spatial extent and a sudden onset. Knowing that the sound source level was fairly low (it was only 226 dB (decibels) @ 3 kHz which is low compared to an operational sonar) would have made the damage mechanism even more puzzling.

- The experimental sonar used in the test, Towed Vertically Directive Source (TVDS) which had the dual 600 Hz and 3 kHz transducers, had been used for the first time in the Mediterranean Sea south of Sicily the year before in June 1995. Previous activated towed array sonar research using different sources on board the same ship included participation in NATO exercises "Dragon Hammer '92" and "Resolute Response '94".

Since the source level of this experimental sonar was only 226 dB @ 3 kHz re. 1 meter, at only 100 meters the received level would drop by 40 dB (to 186 dB). A NATO panel investigated the above stranding and concluded the whales were exposed to 150-160 dB re 1 μPa of low and mid-range frequency sonar. This level is about 66 dB less (more than a million times lower intensity) than the threshold for hearing damage specified by a panel of marine mammal experts.

The idea that a relatively low power sonar could cause a mass stranding of such a large number of whales was very unexpected by the scientific community. Most research had been focused on the possibility of masking signals, interference with mating calls, and similar biological functions. Deep diving marine mammals were species of concern, but very little definitive information was known. In 1995 a comprehensive book on the relation between marine mammals and noise had been published, and it did not even mention strandings.

In 2013, research showed beaked whales were highly sensitive to mid-frequency active sonar. Blue whales have also been shown to flee from the source of mid-frequency sonar, while naval use of mid- and high- frequency side-scan sonar may have caused a mass stranding of dolphins in 2008.

Acoustically Induced Bubble Formation

There was anecdotal evidence from whalers that sonar could panic whales and cause them to surface more frequently making them vulnerable to harpooning. It has also been theorized that military sonar may induce whales to panic and surface too rapidly leading to a form of decompression sickness. In general trauma caused by rapid changes of pressure is known as *barotrauma*. The idea of acoustically enhanced bubble formation was first raised by a paper published in *The Journal of the Acoustical Society of America* in 1996 and again *Nature* in 2003. It reported acute gas-bubble lesions (indicative of decompression sickness) in whales that beached shortly after the start of a military exercise off the Canary Islands in September 2002.

In the Bahamas in 2000, a sonar trial by the United States Navy of transmitters in the frequency range 3–8 kHz at a source level of 223–235 decibels re 1 μPa (scaled to a distance of 1 m) was associated with the beaching of seventeen whales, seven of which were found dead. Environmental groups claimed that some of the beached whales were bleeding from the eyes and ears, which they considered an indication of acoustically-induced trauma. The groups allege that the resulting disorientation may have led to the stranding.

Naval Sonar-linked Incidents

> Worldwide, use of active sonar has been linked to about 50 marine mammal strandings between 1996 and 2006. In all of these occurrences, there were other contributing factors, such as unusual (steep and complex) underwater geography, limited egress routes, and a specific species of marine mammal — beaked whales — that are suspected to be more sensitive to sound than other marine mammals.
>
> *—Rear Admiral Lawrence Rice*

Date	Location	Species and Number	Naval Activity
1963-05	Gulf of Genoa, Italy	Cuvier's beaked whale (15) stranded	Naval maneuvers
1988-11	Canary Islands	Cuvier's beaked whale (12+) Gervais' beaked whale (1) stranded	FLOTA 88 exercise
1989-10	Canary Islands	Cuvier's beaked whale (15+), Gervais' beaked whale (3), Blainville's beaked whale (2) stranded	CANAREX 89 exercise
1991-12	Canary Islands	Cuvier's beaked whale (2) stranded	SINKEX 91 exercise
1996-05-12	Gulf of Kyparissia, Greece	Cuvier's beaked whale (12) stranded	NATO Shallow Water Acoustic Classification exercise
1998-07	Kauai, Hawaii	beaked whale (1), sperm whale (1) stranded	RIMPAC 98 exercise
1999-10	U.S. Virgin Islands and Puerto Rico	Cuvier's beaked whale (4) stranded	COMPTUEX exercise
2000-03-15	Bahamas	Cuvier's beaked whale (9), Blainville's beaked whale (3), beaked whale spp (2), Minke whale (2), Atlantic spotted dolphin (1) stranded	Naval MFA
2000-05-10	Madeira	Cuvier's beaked whale (3) stranded	NATO Linked Seas 2000 and MFA
2002-09	Canary Islands	Cuvier's beaked whale (9), Gervais' beaked whale (1), Blainville's beaked whale (1), beaked whale spp. (3) stranded	Neo Tapon 2002 exercise and MFA
2003-05	Haro Strait, Washington	Harbor porpoise (14), Dall's porpoise (1) Orca avoidance "stampede"	U.S.S. Shoup transiting while using MFA (AN/SQS-53C)
2004-07	Kauai, Hawaii	Melon-headed whale (~200) avoidance "stampede"	RIMPAC 04 exercise with MFA
2004-07-22	Canary Islands	Cuvier's beaked whale (4) stranded	Majestic Eagle 04 exercise
2005-10-25	Marion Bay, Tasmania	Long-finned pilot whales (145) stranded	Two minesweepers using active sonar
2006-01-26	Almería Coast, Spain	Cuvier's beaked whale (4) stranded	HMS Kent using active MF sonar
2008-06	Cornwall, UK	Dolphins (26) stranded	Naval exercise but no ship sonar in use except HF hydrographic sonar on HMS Enterprise

Scientific Attention

Since the 1990s, scientific research has been carried out on the effects of sonar on marine life. This scientific research is reported in peer reviewed journals and at international conferences such as The Effects of Sound on Marine Mammals and The Effects of Noise on Aquatic Life.

A study on the effects of certain sonar frequencies on blue whales was published in 2013. Mid-frequency (1–10 kHz) military sonars have been associated with lethal mass strandings of deep-diving toothed whales, but the effects on endangered baleen whale species were virtually unknown. Controlled exposure experiments, using simulated military sonar and other mid-frequency sounds, measured behavioral responses of tagged blue whales in feeding areas within the Southern California Bight. Despite using source levels orders of magnitude below some operational military systems, the results demonstrated that mid-frequency sound can significantly affect blue whale behavior, especially during deep feeding modes. When a response occurred, behavioral changes varied widely from cessation of deep feeding to increased swimming speed and directed travel away from the sound source. The variability of these behavioral responses was largely influenced by a complex interaction of behavioral state, the type of mid-frequency sound and received sound level. Sonar-induced disruption of feeding and displacement from high-quality prey patches could have significant and previously undocumented impacts on baleen whale foraging ecology, individual fitness and population health.

Court Cases

Since mid-frequency sonar has been correlated with mass cetacean strandings throughout the world's oceans, it has been singled out by some environmentalists as a focus for activism. A lawsuit filed by the Natural Resources Defense Council (NRDC) in Santa Monica, California on 20 October 2005 contended that the U.S. Navy has conducted sonar exercises in violation of several environmental laws, including the National Environmental Policy Act, the Marine Mammal Protection Act, and the Endangered Species Act. Mid-frequency sonar is by far the most common type of active sonar in use by the world's navies, and has been widely deployed since the 1960s.

On November 13, 2007, a United States appeals court restored a ban on the U.S. Navy's use of submarine-hunting sonar in training missions off Southern California until it adopted better safeguards for whales, dolphins and other marine mammals. On 16 January 2008, President George W. Bush exempted the US Navy from the law and argued that naval exercises are crucial to national security. On 4 February 2008, a Federal judge ruled that despite President Bush's decision to exempt it, the Navy must follow environmental laws placing strict limits on mid-frequency sonar. In a 36-page decision, U.S. District Judge Florence-Marie Cooper wrote that the Navy is not "exempted from compliance with the National Environmental Policy Act" and the court injunction creating a 12-nautical-mile (22 km) no-sonar zone off Southern California. On 29 February 2008, a three-judge federal appeals court panel upheld the lower court order requiring the Navy to take precautions during sonar training to minimize harm to marine life. In *Winter v. Natural Resources Defense Council*. the U.S. Supreme Court overturned the circuit court ruling in a 5:4 decision on 12 November 2008.

Mitigation Methods

Environmental impacts of the operation of active sonar are required to be carried out by US law. Procedures for minimising the impact of sonar are developed in each case where there is significant impact.

The impact of underwater sound can be reduced by limiting the sound exposure received by an animal. The maximum sound exposure level recommended by Southall et al. for cetaceans is 215 dB re 1 μPa² s for hearing damage. Maximum sound pressure level for behavioural effects is dependent on context (Southall et al.).

A great deal of the legal and media conflict on this issue has to do with questions of who determines what type of mitigation is sufficient. Coastal commissions, for example, were originally thought to only have legal responsibility for beachfront property, and state waters (three miles into sea). Because active sonar is instrumental to ship defense, mitigation measures that may seem sensible to a civilian agency without any military or scientific background can have disastrous effects on training and readiness. Navies therefore often define their own mitigation requirements.

Examples of mitigation measures include:

1. not operating at nighttime

2. not operating at specific areas of the ocean that are considered sensitive

3. slow ramp-up of intensity of signal to give whales a warning

4. air cover to search for mammals

5. not operating when a mammal is known to be within a certain range

6. onboard observers from civilian groups

7. using fish-finders to look for whales in the vicinity

8. large margins of safety for exposure levels

9. not operating when dolphins are bow-riding

10. operations at less than full power

11. paid teams of veterans to investigate strandings after sonar operation.

References

- Zofia Kielan-Jaworowska, Richard L. Cifelli, Zhe-Xi Luo (2004). "Chapter 7: Eutriconodontans". Mammals from the Age of Dinosaurs: origins, evolution, and structure. New York: Columbia University Press. pp. 216–248. ISBN 0-231-11918-6.

- Kielan-Jaworowska, Zofia; Cifelli, Richard L.; Luo, Zhe-Xi (2004). Mammals from the Age of Dinosaurs: Origins, Evolution, and Structure. New York: Columbia University Press. pp. 441–462. ISBN 978-0-231-50927-5.

- Jefferson, T. A.; Webber, M. A.; Pitman, R. L. (2009). Marine Mammals of the World A Comprehensive Guide to their Identification (1 ed.). London: Academic Press. pp. 7–16. ISBN 978-0-12-383853-7. OCLC 326418543.

- Riedman, M. (1990). The Pinnipeds: Seals, Sea Lions, and Walruses. University of California Press. pp. 11–12. ISBN 978-0-520-06497-3. OCLC 19511610.

- Stirling, Ian; Guravich, Dan (1988). Polar Bears. Ann Arbor, MI: University of Michigan Press. pp. 27–28. ISBN 978-0-472-10100-9. OCLC 757032303.

- Berta, A; Sumich, J. L. (1999). "Exploitation and conservation". Marine Mammals: Evolutionary Biology. San Diego: Academic Press. ISBN 978-0-12-093225-2. OCLC 42467530.

- Kingdon, Jonathan; Happold, David; Butynski, Thomas; Hoffmann, Michael; Happold, Meredith; Kalina,

Jan (2013). Mammals of Africa. 1. New York: Bloomsbury Publishing. p. 211. ISBN 978-1-4081-2251-8. OCLC 822025146.

- Glass, Mogens L.; Wood, Stephen C., eds. (2009). Cardio-Respiratory Control in Vertebrates: Comparative and Evolutionary Aspects. Berlin: Springer. pp. 89–90. doi:10.1007/978-3-540-93984-6. ISBN 978-3-540-93984-9. OCLC 437346699.

- Müller-Schwarze, Dietland & Sun, Lixing (2003). The Beaver: Natural History of a Wetlands Engineer. Cornell University Press. pp. 67–75. ISBN 978-0-8014-4098-4. OCLC 905649783.

- Dalrymple, Byron (1983). North American big-game animals (1 ed.). Stoeger Publishing. p. 84. ISBN 978-08769-1142-6. OCLC 1054473.

- Geist, Valerius (1998). Deer of the World: Their Evolution, Behaviour, and Ecology (1 ed.). Machaniesburg, PA: Stackpole Books. p. 237. ISBN 978-0-8117-0496-0. OCLC 37713037.

- Estes, R. (1992). The Behavior Guide to African Mammals: including hoofed mammals, carnivores, primates. San Francisco: University of California Press. pp. 222–226. ISBN 978-0-520-08085-0. OCLC 19554262.

Various Species of Marine Mammals

The various species of marine mammals are sirenia, manatee, pinniped, sea otter and polar bears. Manatees are large mammals that are fully aquatic and are also sometimes known as sea cows whereas pinnipeds are sometimes known as seals and are semiaquatic marine mammals. The aspects elucidates in this chapter are of vital importance, and provide a better understanding of marine mammals.

Sirenia

Sirenia (commonly referred to as Sea cows) are an order of fully aquatic, herbivorous mammals that inhabit swamps, rivers, estuaries, marine wetlands, and coastal marine waters. Four species are living, in two families and genera. These are the dugong (one species) and manatees (three species). Sirenia also includes Steller's sea cow, extinct since the 18th century, and a number of taxa known only from fossils. The order evolved during the Eocene, more than 50 million years ago.

Sirenia, commonly sirenians, are also referred to by the common name sirens, deriving from the sirens of Greek mythology. This comes from a legend about their discovery, involving lonely sailors mistaking them for mermaids.

"Sea cow" is also the name for a hippopotamus in Afrikaans. In some Germanic languages, the word *Sea* can mean either a body of fresh or salt water, so this follows from the species inhabiting lakes in southern Africa rather than the sea itself.

Description

Sirenians have major aquatic adaptations: the forelimbs have modified into arms used for steering, the tail has modified into a paddle used for propulsion, and the hindlimbs (legs) are but two small remnant bones floating deep in the muscle. They appear fat, but are fusiform, hydrodynamic, and highly muscular. Their skulls are highly modified for taking breaths of air at the water's surface, and dentition is greatly reduced. The skeletal bones of both the manatees and dugong are very dense, which helps to neutralize the buoyancy of their blubber. The manatee appears to have an almost unlimited ability to produce new teeth as the anterior teeth wear down. They have only two teats, located under their forelimbs, similar to elephants. The elephants are thought to be the closest living relatives of the sirenians.

The lungs of sirenians are unlobed. In sirenians, the lungs and diaphragm extend the entire length of the vertebral column. These adaptations help sirenians control their buoyancy and maintain their horizontal position in the water.

Living sirenians grow between 2.5 and 4.0 meters long and can weigh up to 1,500 kg. *Hydrodamalis gigas*, Steller's sea cow, could reach lengths of 8 m.

The three manatee species (family Trichechidae) and the dugong (family Dugongidae) are endangered species. All four are vulnerable to extinction from habitat loss and other negative impacts related to human population growth and coastal development. Steller's sea cow, extinct since 1786, was hunted to extinction by humans. Manatees and dugongs are the only marine mammals classified as herbivores. Unlike the other marine mammals (dolphins, whales, seals, sea lions, sea otters, and walruses), sirenians eat primarily sea grasses and other aquatic vegetation, and have an extremely low metabolism and poor tolerance for especially cold water (the Steller's sea cow, which inhabited the cold waters of the northern Pacific, was an exception). Sirenians have been observed eating dead animals (sea gulls), but their diets are made up primarily of vegetation. Like dolphins and whales, manatees and dugongs are completely aquatic mammals that never leave the water—not even to give birth. These animals have been observed eating grass clippings from homes adjacent to waterways, but in this rare occurrence, only the top portion of the sirenian is lifted out of the water. The combination of these factors means sirenians are restricted to warm, shallow, coastal waters, estuaries, and rivers with healthy ecosystems that support large amounts of seagrass or other vegetation.

The Trichechidae species differ from the Dugongidae in the shape of their skull and tails.

Classification

Pezosiren portelli cast skeleton produced and distributed by Triebold Paleontology Incorporated

The order Sirenia has been placed in the clade Paenungulata, within Afrotheria, grouping it with two other orders of living mammals: Proboscidea, the elephant families, and Hyracoidea, the hyraxes, and two extinct orders, Embrithopoda and Desmostylia.

After Voss, 2014.

† extinct

- ORDER SIRENIA
    - o Family †Prorastomidae
        - Genus †*Pezosiren*
            - †*Pezosiren portelli*
        - Genus †*Prorastomus*

- †*Prorastomus sirenoides*
- Family †Protosirenidae
  - Genus †*Protosiren*
- Family †Archaeosirenidae
  - Genus †*Eosiren*
- Family †Eotheroididae
  - Genus †*Eotheroides*
- Family †Prototheriidae
  - Genus †*Prototherium*
- Family Dugongidae
  - Genus †*Nanosiren*
  - Genus †*Sirenotherium*
  - Subfamily Dugonginae
    - Genus *Dugong*
      - *Dugong dugon*, dugong
  - Subfamily †Hydrodamalinae
    - Genus †*Dusisiren*
    - Genus †*Hydrodamalis*
      - †*Hydrodamalis cuestae*
      - †*Hydrodamalis gigas*, Steller's sea cow
- Family Trichechidae
  - Subfamily †Miosireninae
    - Genus †*Anomotherium*
    - Genus †*Miosiren*
    - Genus †*Prohalicore*
  - Subfamily Trichechinae
    - Genus †*Potamosiren*
    - Genus *Trichechus*
      - *T. manatus*, West Indian manatee
        - *T. m. manatus*, Antillean manatee

- - *T. m. latirostris*, Florida manatee
  - *T. senegalensis*, African manatee
  - *T. inunguis*, Amazonian manatee
  - *T. "pygmaeus"*, dwarf manatee
  - Genus †*Ribodon*

Manatee

Manatees (family Trichechidae, genus *Trichechus*) are large, fully aquatic, mostly herbivorous marine mammals sometimes known as sea cows. There are three accepted living species of Trichechidae, representing three of the four living species in the order Sirenia: the Amazonian manatee (*Trichechus inunguis*), the West Indian manatee (*Trichechus manatus*), and the West African manatee (*Trichechus senegalensis*). They measure up to 4.0 metres (13.1 ft) long, weigh as much as 590 kilograms (1,300 lb), and have paddle-like flippers. The name *manatí* comes from the Spanish "manatí", derived from the Caribbean word sometimes cited as "manattouï". The etymology is dubious, with connections having been made to Latin "manus" (hand), and to a word used by the Taíno, a pre-Columbian people of the Caribbean, meaning "breast". Manatees are occasionally called sea cows, as they are slow plant-eaters, peaceful and similar to cows on land. They often graze on water plants in tropical seas.

Taxonomy

Manatees comprise three of the four living species in the order Sirenia. The fourth is the Eastern Hemisphere's dugong. The Sirenia are thought to have evolved from four-legged land mammals over 60 million years ago, with the closest living relatives being the Proboscidea (elephants) and Hyracoidea (hyraxes).

The Amazonian's hair color is brownish gray, and they have thick wrinkled skin often with coarse hair, or "whiskers". Photos are rare; although very little is known about this species, scientists think they are similar to West Indian manatees.

Description

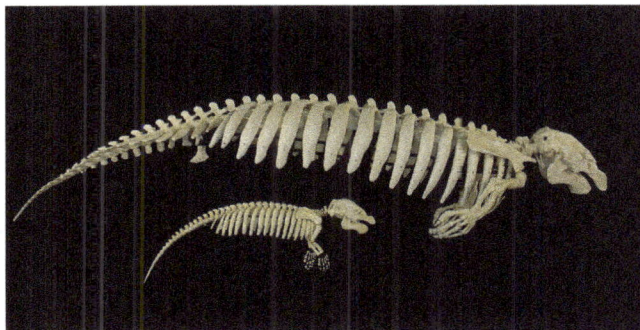

A skeleton of a manatee and calf, on display at The Museum of Osteology, Oklahoma City, Oklahoma

Skull of a West Indian manatee on display at The Museum of Osteology, Oklahoma City, Oklahoma.

Manatees have a mass of 400 to 550 kilograms (880 to 1,210 lb), and mean length of 2.8 to 3.0 metres (9.2 to 9.8 ft), with maxima of 3.6 metres (12 ft) and 1,775 kilograms (3,913 lb) seen (the females tend to be larger and heavier). When born, baby manatees have an average mass of 30 kilograms (66 lb). They have a large, flexible, prehensile upper lip. They use the lip to gather food and eat, as well as using it for social interactions and communications. Manatees have shorter snouts than their fellow sirenians, the dugongs. Their small, widely spaced eyes have eyelids that close in a circular manner. The adults have no incisor or canine teeth, just a set of cheek teeth, which are not clearly differentiated into molars and premolars. These teeth are continuously replaced throughout life, with new teeth growing at the rear as older teeth fall out from farther forward in the mouth. This process is known as polyphyodonty and amongst the other mammals, only occurs in the kangaroo and elephant. At any given time, a manatee typically has no more than six teeth in each jaw of its mouth. Its tail is paddle-shaped, and is the clearest visible difference between manatees and dugongs; a dugong tail is fluked, similar in shape to a that of a whale. Females have two teats, one under each flipper, a characteristic that was used to make early links between the manatee and elephants.

Manatees are unusual amongst mammals in possessing just six cervical vertebrae, which may be due to mutations in the homeotic genes. All other mammals have seven cervical vertebrae, other than the two-toed and three-toed sloths.

Like horses, they have a simple stomach, but a large cecum, in which they can digest tough plant matter. In general, their intestines have a typical length of about 45 meters, which is unusually long for animals of their size. Manatees produce enormous amounts of gas, which contributes to their barrel-shape, to aid in the digestion of their food.

Manatees are the only animal known to have a vascularized cornea.

Behavior

Apart from mothers with their young, or males following a receptive female, manatees are generally solitary animals. Manatees spend approximately 50% of the day sleeping submerged, surfacing for air regularly at intervals of less than 20 minutes. The remainder of the time is mostly spent grazing in shallow waters at depths of 1–2 metres (3.3–6.6 ft). The Florida subspecies (*T. m. latirostris*) has been known to live up to 60 years.

Locomotion

Generally, manatees swim at about 5 to 8 kilometres per hour (3 to 5 mph). However, they have been known to swim at up to 30 kilometres per hour (20 mph) in short bursts.

Intelligence and Learning

Manatee postures in captivity.

Manatees are capable of understanding discrimination tasks and show signs of complex associative learning. They also have good long-term memory. They demonstrate discrimination and task-learning abilities similar to dolphins and pinnipeds in acoustic and visual studies.

Reproduction

Manatees typically breed once every two years; generally only a single calf is born. Gestation lasts about 12 months and a further 12 to 18 months to wean the calf.

Communication

Manatees emit a wide range of sounds used in communication, especially between cows and their calves. Adults communicate to maintain contact and during sexual and play behaviors. Taste and smell, in addition to sight, sound, and touch, may also be forms of communication.

Diet

Manatees are herbivores and eat over 60 different freshwater (e.g. floating hyacinth, pickerel weed, alligator weed, water lettuce, hydrilla, water celery, musk grass, mangrove leaves) and salt-water plants (e.g. sea grasses, shoal grass, manatee grass, turtle grass, widgeon grass, sea clover, and marine algae). Using their divided upper lip, an adult manatee will commonly eat up to 10%-15% of their body weight (about 50 kg) per day. Consuming such an amount requires the manatee to graze for up to seven hours a day. Manatees have been known to eat small amounts of fish from nets.

Feeding Behavior

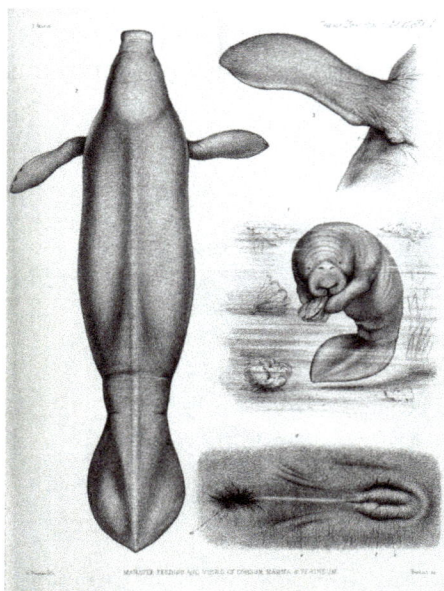

Manatee plate

Manatees use their flippers to "walk" along the bottom whilst they dig for plants and roots in the substrate. When plants are detected, the flippers are used to scoop the vegetation toward the manatee's lips. The manatee has prehensile lips; the upper lip pad is split into left and right sides which can move independently. The lips use seven muscles to manipulate and tear at plants. Manatees use their lips and front flippers to move the plants into the mouth. The manatee does not have front teeth, however, behind the lips, on the roof of the mouth, there are dense, ridged pads. These horny ridges, and the manatee's lower jaw, tear through ingested plant material.

Dentition

Manatees have four rows of teeth. There are 6 to 8 high-crowned, open-rooted molars located along each side of the upper and lower jaw giving a total of 24 to 32 flat, rough-textured teeth. Eating gritty vegetation abrades the teeth, particularly the enamel crown, however, research indicates that the enamel structure in manatee molars is weak. To compensate for this, manatee teeth are continually replaced. When anterior molars wear down, they are shed. Posterior molars erupt at the back of the row and slowly move forward to replace these like enamel crowns on a conveyor belt. This process continues throughout the manatee's lifetime. The rate at which the teeth migrate forward depends on how quickly the anterior teeth abrade. Some studies indicate that the rate is about 1 cm/month although other studies indicate 0.1 cm/month. This process of teeth being continually replaced is known as polyphyodonty and amongst other mammals, only occurs in elephants and kangaroos.

Ecology

Range and Habitat

Manatees inhabit the shallow, marshy coastal areas and rivers of the Caribbean Sea and the Gulf of Mexico (*T. manatus*, West Indian manatee), the Amazon Basin (*T. inunguis*, Amazonian manatee), and West Africa (*T. senegalensis*, West African manatee).

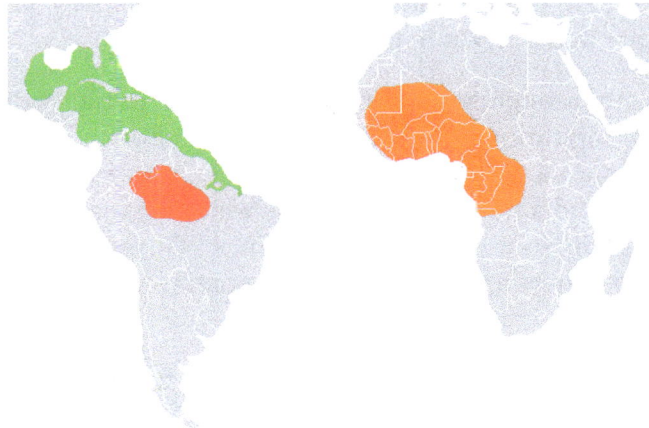

Approximate distribution of *Trichechus*; *T. manatus* in green; *T. inunguis* in red; *T. senegalenis* in orange

West Indian manatees prefer warmer temperatures and are known to congregate in shallow waters. They frequently migrate through brackish water estuaries to freshwater springs. They cannot survive below 15 °C (60 °F). Their natural source for warmth during winter is warm, spring-fed rivers.

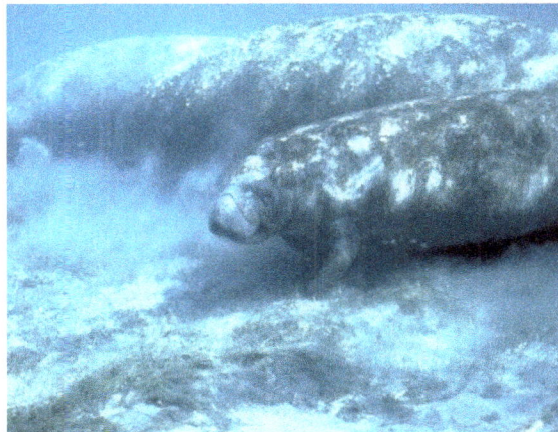

A group of three manatees

West Indian

The coast of the state of Georgia is usually the northernmost range of the West Indian manatees because their low metabolic rate does not protect them in cold water. Prolonged exposure to water temperatures below 20 °C (68 °F) can bring about "cold stress syndrome" and death.

Florida manatees can move freely between salinity extremes.

Manatees have been seen as far north as Cape Cod, and in 1995 and again in 2006, one was seen in New York City and Rhode Island's Narragansett Bay. A manatee was spotted in the Wolf River harbor near the Mississippi River in downtown Memphis in 2006, though it was later found dead 10 miles downriver in McKellar Lake.

The West Indian manatee migrates into Florida rivers, such as the Crystal, the Homosassa, and the Chassahowitzka Rivers. The headsprings of these rivers maintain a 22 °C (72 °F) temperature all year. During November to March, about 400 West Indian manatees (according to the National Wildlife Refuge) congregate in the rivers in Citrus County, Florida.

During winter, manatees often congregate near the warm-water outflows of power plants along the coast of Florida instead of migrating south as they once did. Some conservations are concerned these manatees have become too reliant on these artificially warmed areas. The U.S. Fish and Wildlife Service is trying to find a new way to heat the water for manatees that are dependent on plants that have closed. The main water treatment plant in Guyana has four manatees that keep storage canals clear of weeds; there are also some in the ponds of the National Park in Georgetown, Guyana.

Studies suggest Florida manatees must have some access to fresh water for proper regulation of water and salts within their bodies.

Accurate population estimates of the Florida manatee (*T. manatus*) are difficult. They have been called scientifically weak due to widely varying counts from year to year, some areas showing increases, others decreases and little strong evidence of increases except in two areas. Manatee counts are highly variable without an accurate way to estimate numbers: In Florida in 1996, a winter survey found 2,639 manatees, in 1997, a January survey found 2,229, and a February survey found 1,706. A statewide synoptic survey in January 2010 found 5,067 manatees living in Florida, which was a new record count.

As of January 2016, the USFWS estimates the range-wide manatee population to be at least 13,000, with at least 6,300 in Florida.

Population viability studies conducted in 1997 found that decreasing adult survival and eventual extinction is a probable future outcome for Florida manatees, without additional protection. U.S. Fish and Wildlife Service proposed downgrading the manatee's status from endangered to threatened in January 2016 after over 40 years on the endangered species list.

Fossil remains of Florida manatee ancestors date back about 45 million years.

Amazonian

The freshwater Amazonian manatee (*T. inunguis*) inhabits the Amazon River and its tributaries, and never ventures into salt water.

West African

They are found in coastal marine and estuarine habitats, and in freshwater river systems along the west coast of Africa from the Senegal River south to the Kwanza River in Angola. They live as far upriver on the Niger River as Koulikoro in Mali, 2,000 km from the coast.

Predation

Overall, predation does not present a significant threat to the survival of any manatee species.

Relation to Humans

Young manatees can be curious; this individual is inspecting a kayak

Antillean manatee

Threats

The main causes of death for manatees are human-related issues, such as habitat destruction and human objects. Natural causes of death include adverse temperatures and disease.

Ship Strikes

Their slow-moving, curious nature, coupled with dense coastal development, has led to many violent collisions with propeller-driven boats and ships, leading frequently to maiming, disfigurement, and even death. As a result, a large proportion of manatees exhibit spiral cutting propeller scars on their backs, usually caused by larger vessels that do not have skegs in front of the propellers like the smaller outboard and inboard-outboard recreational boats have. They are now even identified by humans based on their scar patterns. Many manatees have been cut in half by large vessels like ships and tug boats, even in the highly populated lower St. Johns River's narrow channels. Some are concerned that the current situation is inhumane, with upwards of 50 scars and disfigurements from vessel strikes on a single manatee. Often, the cuts lead to infections, which can prove fatal. Internal injuries stemming from being trapped between hulls and docks and impacts have also been fatal. Recent testing shows that manatees may be able to hear speed boats and other watercraft approaching, due to the frequency the boat makes. However, a manatee may not be able to hear the approaching boats when they are performing day-to-day activities or distractions. The manatee has a tested frequency range of 8 kilohertz to 32 kilohertz.

Manatees hear on a higher frequency than would be expected for such large marine mammals. Many large boats emit very low frequencies, which confuse the manatee and explain their lack of awareness around boats. The Lloyd's mirror effect results in low frequency propeller sounds not being discernible near the surface, where most accidents occur. Research indicates that when a boat has a higher frequency the manatees rapidly swim away from danger.

In 2003, a population model was released by the United States Geological Survey that predicted an extremely grave situation confronting the manatee in both the Southwest and Atlantic regions where the vast majority of manatees are found. It states,

"In the absence of any new management action, that is, if boat mortality rates continue to increase at the rates observed since 1992, the situation in the Atlantic and Southwest regions is dire, with no chance of meeting recovery criteria within 100 years."

"Hurricanes, cold stress, red tide poisoning and a variety of other maladies threaten manatees, but by far their greatest danger is from watercraft strikes, which account for about a quarter of Florida manatee deaths," said study curator John Jett.

According to marine mammal veterinarians:

"The severity of mutilations for some of these individuals can be astounding – including long term survivors with completely severed tails, major tail mutilations, and multiple disfiguring dorsal lacerations. These injuries not only cause gruesome wounds, but may also impact population processes by reducing calf production (and survival) in wounded females – observations also speak to the likely pain and suffering endured". In an example, they cited one case study of a small calf "with a severe dorsal mutilation trailing a decomposing piece of dermis and muscle as it continued to accompany and nurse from its mother...by age 2 its dorsum was grossly deformed and included a large protruding rib fragment visible."

These veterinarians go on to state:

"the overwhelming documentation of gruesome wounding of manatees leaves no room for denial. Minimization of this injury is *explicit* in the Recovery Plan, several state statutes, and federal laws, and *implicit* in our society's ethical and moral standards."

In 2009, of the 429 Florida manatees recorded dead, 97 were killed by commercial and recreational vessels, which broke the earlier record number of 95 set in 2002.

Red Tide

Another cause of manatee deaths are red tides, a term used for the proliferation, or "blooms", of the microscopic marine algae, *Karenia brevis*. This dinoflagellate produces brevetoxins that can have toxic effects on the central nervous system of animals.

In 1996, a red tide was responsible for 151 manatee deaths. The bloom was present from early March to the end of April and killed approximately 15% of the known population of manatees along South Florida's western coast. Other blooms in 1982 and 2005 resulted in 37 and 44 deaths, respectively.

Additional Threats

Manatees occasionally ingest fishing gear (hooks, metal weights, etc.) while feeding. These foreign materials do not appear to harm manatees, except for monofilament line or string, which can block a manatee's digestive system and slowly kill it.

Manatees can also be crushed in water control structures (navigation locks, floodgates, etc.), drown in pipes and culverts, and are occasionally killed by entanglement in fishing gear, primarily crab pot float lines.

While humans are allowed to swim with manatees in one area of Florida, there have been numerous charges of people harassing and disturbing the manatees. According to the United States Fish and Wildlife Service, approximately 99 manatee deaths each year are related to human activities. In January 2016, there were 43 manatee deaths in Florida alone.

Conservation

All three species of manatee are listed by the World Conservation Union as vulnerable to extinction.

It is illegal under federal and Florida law to injure or harm a manatee. They are classified as "endangered" by both the state and the federal governments.

The MV *Freedom Star* and MV *Liberty Star*, ships used by NASA to tow space shuttle solid rocket boosters back to Kennedy Space Center, are propelled only by water jets to protect the endangered manatee population that inhabits regions of the Banana River where the ships are based.

Brazil outlawed hunting in 1973 in an effort to preserve the species. Deaths by boat strikes are still common.

In January 2016, the U.S. Fish and Wildlife Service proposed that the West Indian Manatee be reclassified from an "endangered" status to "threatened" as improvements to habitat conditions, population growth and reductions of threats have all increased. The proposal will not affect current federal protections.

As of February 2016, 6,250 manatees were reported swimming in Florida's springs.

Hunting

Trichechus sp.

Manatees were traditionally hunted by indigenous Caribbean people. When Christopher Columbus arrived in the region, hunting was already an established trade, although this is less common today.

The primary hunting method was for the hunter to approach in a dugout canoe, offering bait to attract it close enough to temporarily stun it with a blow near the head from an oar-like pole. Many times the creature would flip over, leaving it vulnerable to further attacks.

From manatee hides, Native Americans made war shields, canoes, and shoes, though manatees were predominantly hunted for their abundant meat.

Later, manatees were hunted for their bones, which were used to make "special potions". Until the 1800s, museums paid as much as $100 for bones or hides. Though hunting was banned in 1893, poaching continues today.

Captivity

A manatee at SeaWorld, Florida

The oldest manatee in captivity is Snooty, at the South Florida Museum's Parker Manatee Aquarium in Bradenton, Florida. Born at the Miami Aquarium and Tackle Company on July 21, 1948, Snooty was one of the first recorded captive manatee births. Raised entirely in captivity, Snooty will never be released into the wild. As such he is the only manatee at the Aquarium, and one of a select few captive manatees in the United States, that is allowed to interact with human handlers. This makes him uniquely suitable to manatee research and education.

Manatees can also be viewed in a number of European zoos, such as the Tierpark in Berlin, the Nuremberg Zoo, in ZooParc de Beauval in France and in the Aquarium of Genoa in Italy. The River Safari at Singapore features seven of them. They are also included in the plans for a new National Wildlife Conservation Park in Bristol, England, whose first exhibit is due to open in summer 2013 with the manatees as an addition as early as 2015.

Culture

The manatee has been linked to folklore on mermaids. Native Americans ground the bones to treat asthma and earache. In West African folklore, they were considered sacred and thought to have been once human. Killing one was taboo and required penance.

Manatees were featured in the "Cartoon Wars Part II" episode of *South Park*, as the creative force behind the television show *Family Guy*. The manatees were shown to be living in a tank at FOX Studios which was filled with "idea balls." The manatees randomly selected the idea balls to make the jokes for the show. They are also revealed as being "the only animal unmoved by terrorist threats."

Pinniped

Pinnipeds (from Latin *pinna* "fin" and *pes, pedis* "foot"), commonly known as seals, are a widely distributed and diverse clade of carnivorous, fin-footed, semiaquatic marine mammals. They comprise the extant families Odobenidae (whose only living member is the walrus), Otariidae (the eared seals: sea lions and fur seals), and Phocidae (the earless seals, or true seals). There are 33 extant species of pinnipeds, and more than 50 extinct species have been described from fossils. While seals were historically thought to have descended from two ancestral lines, molecular evidence supports them as a monophyletic lineage (descended from one ancestral line). Pinnipeds belong to the order Carnivora and their closest living relatives are bears and musteloids (weasels, raccoons, skunks and red pandas), having diverged about 50 million years ago.

Seals range in size from the 1 m (3 ft 3 in) and 45 kg (99 lb) Baikal seal to the 5 m (16 ft) and 3,200 kg (7,100 lb) southern elephant seal, which is also the largest carnivoran.[b] Several species exhibit sexual dimorphism. They have streamlined bodies and four limbs that are modified into flippers. Though not as fast in the water as dolphins, seals are more flexible and agile. Otariids use their front limbs primarily to propel themselves through the water, while phocids and walruses use their hind limbs. Otariids and walruses have hind limbs that can be pulled under the body and used as legs on land. By comparison, terrestrial locomotion by phocids is more cumbersome. Otariids have visible external ears, while phocids and walruses lack these. Pinnipeds have well-developed senses—their eyesight and hearing are adapted for both air and water, and they have an advanced tactile system in their whiskers or vibrissae. Some species are well adapted for diving to great depths. They have a layer of fat, or blubber, under the skin to keep warm in the cold water, and, other than the walrus, all species are covered in fur.

Although pinnipeds are widespread, most species prefer the colder waters of the Northern and Southern Hemispheres. They spend most of their lives in the water, but come ashore to mate, give birth, molt or escape from predators, like sharks and killer whales. They feed largely on fish and marine invertebrates; but a few, like the leopard seal, feed on large vertebrates, such as penguins and other seals. Walruses are specialized for feeding on bottom-dwelling mollusks. Male pinnipeds typically mate with more than one female (polygyny), although the degree of polygyny varies with the species. The males of land-breeding species tend to mate with a greater number of females than those of ice- or water-breeding species. Male pinniped strategies for reproductive success vary between defending females, defending territories that attract females and performing ritual displays or lek mating. Pups are typically born in the spring and summer months and females bear almost all the responsibility for raising them. Mothers of some species fast and nurse their young for a relatively short period of time while others take foraging trips at sea between nursing bouts. Walruses are known to nurse their young while at sea. Seals produce a number of vocalizations, notably the barks of California sea lions, the gong-like calls of walruses and the complex songs of Weddell seals.

The meat, blubber and fur coats of pinnipeds have traditionally been used by indigenous peoples of the Arctic. Seals have been depicted in various cultures worldwide. They are commonly kept in captivity and are even sometimes trained to perform tricks and tasks. Once relentlessly hunted by commercial industries for their products, seals and walruses are now protected by international law. The Japanese sea lion and the Caribbean monk seal have become extinct in the past century, while the Mediterranean monk seal and Hawaiian monk seal are ranked Critically Endangered by the International Union for Conservation of Nature. Besides hunting, pinnipeds also face threats from accidental trapping, marine pollution, and conflicts with local people.

The German naturalist Johann Karl Wilhelm Illiger was the first to recognize the pinnipeds as a distinct taxonomic unit; in 1811 he gave the name Pinnipedia to both a family and an order. American zoologist Joel Asaph Allen reviewed the world's pinnipeds in an 1880 monograph, *History of North*

American pinnipeds, a monograph of the walruses, sea-lions, sea-bears and seals of North America. In this publication, he traced the history of names, gave keys to families and genera, described North American species and provided synopses of species in other parts of the world. In 1989, Annalisa Berta and colleagues proposed the unranked clade Pinnipedimorpha to contain the fossil genus *Enaliarctos* and modern seals as a sister group. Pinnipeds belong to the order Carnivora and the suborder Caniformia (known as dog-like carnivorans). Pinnipedia was historically considered its own suborder under Carnivora. Of the three extant families, the Otariidae and Odobenidae are grouped in the superfamily Otarioidea, while the Phocidae belong to the superfamily Phocoidea.

Otariids are also known as eared seals due to the presence of pinnae. These animals rely on their well-developed fore-flippers to propel themselves through the water. They can also turn their hind-flippers forward and "walk" on land. The anterior end of an otariid's frontal bones extends between the nasal bones, and the supraorbital foramen is large and flat horizontally. The supraspinatous fossas are divided by a "secondary spine" and the bronchi are divided anteriorly. Otariids consist of two types: sea lions and fur seals. Sea lions are distinguished by their rounder snouts and shorter, rougher pelage, while fur seals have more pointed snouts, longer fore-flippers and thicker fur coats that include an undercoat and guard hairs. The former also tend to be larger than the latter. Five genera and seven species (one now extinct) of sea lion are known to exist, while two genera and nine species of fur seal exist. While sea lions and fur seals have historically been considered separate subfamilies (Otariinae and Arctocephalinae respectively), a 2001 genetic study found that the northern fur seal is more closely related to several sea lion species. This is supported by a 2006 molecular study that also found that the Australian sea lion and New Zealand sea lion are more closely related to *Arctocephalus* than to other sea lions.

Odobenidae consists of only one living member: the modern walrus. This animal is easily distinguished from other extant pinnipeds by its larger size (exceeded only by the elephant seals), nearly hairless skin and long upper canines, known as tusks. Like otariids, walruses are capable of turning their hind-flippers forward and can walk on land. When moving in water, the walrus relies on its hind-flippers for locomotion, while its fore-flippers are used for steering. In addition, the walrus lacks external ear flaps. Walruses have pterygoid bones that are broad and thick, frontal bones that are V-shaped at the anterior end and calcaneuses with pronounced tuberosity in the middle.

Phocids are known as true or "earless" seals. These animals lack external ear flaps and are incapable of turning their hind-flippers forward, which makes them more cumbersome on land. In water, true seals swim by moving their hind-flippers and lower body from side to side. Phocids have thickened mastoids, enlarged entotympanic bones, everted pelvic bones and massive ankle bones. They also lack supraorbital processes on the frontal and have underdeveloped calcaneal tubers. A 2006 molecular study supports the division of phocids into two monophyletic subfamilies: Monachinae, which consists of *Mirounga*, Monachini and Lobodontini; and Phocinae, which includes *Pusa, Phoca, Halichoerus, Histriophoca, Pagophilus, Erignathus* and *Cystophora*.

In a 2012 review of pinniped taxonomy, Berta and Morgan Churchill suggested that, based on morphological and genetic criteria, there are 33 extant species and 29 subspecies of pinnipeds, although five of the latter lack sufficient support to be conclusively considered subspecies. They recommend that the genus *Arctocephalus* be limited to *Arctocephalus pusillus*, and they resurrected the name *Arctophoca* for several species and subspecies formerly placed in *Arctocephalus*. More than 50 fossil species have been described.

Evolutionary History

Restoration of *Puijila darwini*

One popular hypothesis suggested that pinnipeds are diphyletic (descended from two ancestral lines), with walruses and otariids sharing a recent common ancestor with bears and phocids sharing one with Musteloidea. However, morphological and molecular evidence support a monophyletic origin. Nevertheless, there is some dispute as to whether pinnipeds are more closely related to bears or musteloids, as some studies support the former theory and others the latter. Pinnipeds split from other caniforms 50 million years ago (mya) during the Eocene. Their evolutionary link to terrestrial mammals was unknown until the 2007 discovery of *Puijila darwini* in early Miocene deposits in Nunavut, Canada. Like a modern otter, *Puijila* had a long tail, short limbs and webbed feet instead of flippers. However, its limbs and shoulders were more robust and *Puijila* likely had been a quadrupedal swimmer—retaining a form of aquatic locomotion that give rise to the major swimming types employed by modern pinnipeds. The researchers who found *Puijila* placed it in a clade with *Potamotherium* (traditionally considered a mustelid) and *Enaliarctos*. Of the three, *Puijila* was the least specialized for aquatic life. The discovery of *Puijila* in a lake deposit suggests that pinniped evolution went through a freshwater transitional phase.

Fossil of *Enaliarctos*

Enaliarctos, a fossil species of late Oligocene/early Miocene (24–22 mya) California, closely resembled modern pinnipeds; it was adapted to an aquatic life with a flexible spine, and limbs modified into flippers. Its teeth were adapted for shearing (like terrestrial carnivorans), and it may have stayed near shore more often than its extant relatives. *Enaliarctos* was capable of swimming with both the fore-flippers and hind-flippers, but it may have been more specialized as a fore-flipper swimmer. One species, *Enaliarctos emlongi*, exhibited notable sexual dimorphism, suggesting that this physical characteristic may have been an important driver of pinniped evolution. A closer relative of extant pinnipeds was *Pteroarctos*, which lived in Oregon 19–15 mya. As in modern seals, *Pteroarctos* had an orbital wall that was not limited by certain facial bones (like the jugal or lacrimal bone), but was mostly shaped by the maxilla.

The lineages of Otariidae and Odobenidae split almost 28 mya. Otariids originated in the North Pacific. The earliest fossil *Pithanotaria*, found in California, is dated to 11 mya. The *Callorhinus* lineage split earlier at 16 mya. *Zalophus*, *Eumetopias* and *Otaria* diverged next, with the latter colonizing the coast of South America. Most of the other otariids diversified in the Southern Hemisphere. The earliest fossils of Odobenidae—*Prototaria* of Japan and *Proneotherium* of Oregon—date to 18–16 mya. These primitive walruses had much shorter canines and lived on a fish diet rather than a specialized mollusk diet like the modern walrus. Odobenids further diversified in the middle and late Miocene. Several species had enlarged upper and lower canines. The genera *Valenictus* and *Odobenus* developed elongated tusks. The lineage of the modern walrus may have spread from the North Pacific to the Caribbean (via the Central American Seaway) 8–5 mya and subsequently made it to the North Atlantic and returned to the North Pacific via the Arctic 1 mya. Alternatively, this lineage may have spread from the North Pacific to the Arctic and subsequently the North Atlantic during the Pleistocene.

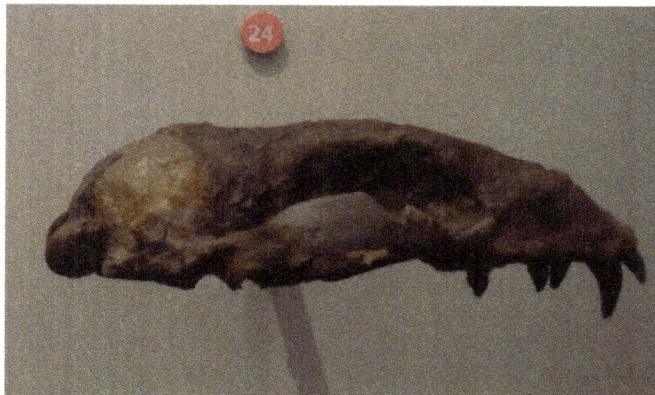

Fossil skull cast of *Desmatophoca oregonensis* from the extinct Desmatophocidae

The ancestors of the Otarioidea and Phocoidea diverged 33 mya. The Phocidae are likely to have descended from the extinct family Desmatophocidae in the North Atlantic. Desmatophocids lived 23–10 mya and had elongated skulls, fairly large eyes, cheekbones connected by a mortised structure and rounded cheek teeth. They also were sexually dimorphic and may have been capable of propelling themselves with both the foreflippers and hindflippers.

Phocids are known to have existed for at least 15 mya, and molecular evidence supports a divergence of the Monachinae and Phocinae lineages 22 mya. The fossil monachine *Monotherium* and phocine *Leptophoca* were found in southeastern North America. The deep split between the lineages of *Erignathus* and *Cystophora* 17 mya suggests that the phocines migrated eastward and northward from the North Atlantic. The genera *Phoca* and *Pusa* could have arisen when a phocine lineage traveled from the Paratethys Sea to the Arctic Basin and subsequently went eastward. The ancestor of the Baikal seal migrated into Lake Baikal from the Arctic (via the Siberian ice sheet) and became isolated there. The Caspian seal's ancestor became isolated as the Paratethys shrank, leaving the animal in a small remnant sea, the Caspian Sea. The monochines diversified southward. *Monachus* emerged in the Mediterranean and migrated to the Caribbean and then the central North Pacific. The two extant elephant seal species diverged close to 4 mya after the Panamanian isthmus was formed. The lobodontine lineage emerged around 9 mya and colonized the southern ocean in response to glaciation.

Anatomy and Physiology

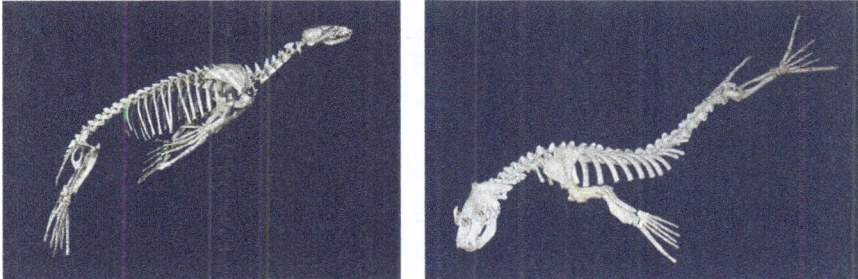

Skeleton of California sea lion (above) and southern elephant seal

Pinnipeds have streamlined, spindle-shaped bodies with reduced or non-existent external ear flaps, rounded heads, flexible necks, limbs modified into flippers, and small tails. Pinniped skulls have large eye orbits, short snouts and a constricted interorbital region. They are unique among carnivorans in that their orbital walls are significantly shaped by the maxilla and are not limited by certain facial bones. Compared to other carnivorans, their teeth tend to be fewer in number (especially incisors and back molars), are pointed and cone-shaped, and lack carnassials. The walrus has unique upper canines that are elongated into tusks. The mammary glands and genitals of pinnipeds can retract into the body.

Pinnipeds range in size from the 1 m (3 ft 3 in) and 45 kg (99 lb) Baikal seal to the 5 m (16 ft) and 3,200 kg (7,100 lb) southern elephant seal. Overall, they tend to be larger than other carnivorans; the southern elephant seal is the largest carnivoran. Several species have male-biased sexual dimorphism that correlates with the degree of polygyny in a species: highly polygynous species like elephant seals are extremely sexually dimorphic, while less polygynous species have males and females that are closer in size. In lobodontine seals, females are slightly larger than males. Males of sexually dimorphic species also tend to have secondary sex characteristics, such as the prominent proboscis of elephant seals, the inflatable red nasal membrane of hooded seals and the thick necks and manes of otariids. Despite a correlation between size dimorphism and the degree of polygyny, some evidence suggests that size differences between the sexes originated due to ecological differences and prior to the development of polygyny.

Male and female South American sea lions, showing sexual dimorphism

Almost all pinnipeds have fur coats, the exception being the walrus, which is only sparsely covered. Even some fully furred species (particularly sea lions) are less haired than most land mammals. In species that live on ice, young pups have thicker coats than adults. The individual hairs on the coat, known collectively as lanugo, can trap heat from sunlight and keep the pup warm. Pinnipeds are typically countershaded, and are darker colored dorsally and lighter colored ventrally, which serves to eliminate shadows caused by light shining over the ocean water. The pure white fur of harp seal pups conceals them in their Arctic environment. Some species, such as ribbon seals, ringed seals and leopard seals, have patterns of contrasting light and dark coloration. All fully furred species molt; phocids molt once a year, while otariids gradually molt all year. Seals have a layer of subcutaneous fat known as blubber that is particularly thick in phocids and walruses. Blubber serves both to keep the animals warm and to provide energy and nourishment when they are fasting. It can constitute as much as 50% of a pinniped's body weight. Pups are born with only a thin layer of blubber, but some species compensate for this with thick lanugos.

Pinnipeds have a simple stomach that is similar in structure to terrestrial carnivores. Most species have neither a cecum nor a clear demarcation between the small and large intestines; the large intestine is comparatively short and only slightly wider than the latter. Small intestine lengths range from 8 (California sea lion) to 25 times (elephant seal) the body length. The length of the intestine may be an adaptation to frequent deep diving, as the increased volume of the digestive tract serves as an extended storage compartment for partially digested food during submersion. Pinnipeds do not have an appendix. As in most marine mammals, the kidneys are divided into small lobes and can effectively absorb water and filter out excess salt.

Locomotion

Harbor seal (above) and California sea lion swimming. The former swims with its hind-flippers, the latter with its fore-flippers.

Pinnipeds have two pairs of flippers on the front and back, the fore-flippers and hind-flippers. The elbows and ankles are enclosed within the body. Pinnipeds tend to be slower swimmers than cetaceans, typically cruising at 5–15 kn (9–28 km/h; 6–17 mph) compared to around 20 kn (37 km/h; 23 mph) for several species of dolphin. Seals are more agile and flexible, and some otariids, such as the California sea lion, are capable of bending their necks backwards far enough to reach their hind-flippers, allowing them to make dorsal turns. Pinnipeds have several adaptions for reducing drag. In addition to their streamlined bodies, they have smooth networks of muscle bundles in their skin that may increase laminar flow and make it easier for them to slip through water. They also lack arrector pili, so their fur can be streamlined as they swim.

When swimming, otariids rely on their fore-flippers for locomotion in a wing-like manner similar to penguins and sea turtles. Fore-flipper movement is not continuous, and the animal glides between each stroke. Compared to terrestrial carnivorans, the fore-limbs of otariids are reduced in length, which gives the locomotor muscles at the shoulder and elbow joints greater mechanical advantage; the hind-flippers serve as stabilizers. Phocids and walruses swim by moving their hind-flippers and lower body from side to side, while their fore-flippers are mainly used for steering. Some species leap out of the water, which may allow then to travel faster. In addition, sea lions are known to "ride" waves, which probably helps them decrease their energy usage.

Pinnipeds can move around on land, though not as well as terrestrial animals. Otariids and walruses are capable of turning their hind-flippers forward and under the body so they can "walk" on all fours. The fore-flippers move in a transverse, rather than a sagittal fashion. Otariids rely on the movements of their heads and necks more than their hind-flippers during terrestrial locomotion. By swinging their heads and necks, otariids create momentum while they are moving. Sea lions have been recorded climbing up flights of stairs. Phocids are less agile on land. They cannot pull their hind-flippers forward, and move on land by lunging, bouncing and wiggling while their fore-flippers keep them balanced. Some species use their fore-flippers to pull themselves forward. Terrestrial locomotion is easier for phocids on ice, as they can sled along.

Senses

Frontal view of brown fur seal head

The eyes of pinnipeds are relatively large for their size and are positioned near the front of the head. One exception is the walrus, whose smaller eyes are located on the sides of its head. This is because it feeds on immobile bottom dwelling mollusks and hence does not need acute vision. A seal's eye is adapted for seeing both underwater and in air. The lens is mostly spherical, and much of the retina is equidistant from the lens center. The cornea has a flattened center where refraction is nearly equal in both water and air. Pinnipeds also have very muscular and vascularized irises.

The well-developed dilator muscle gives the animals a great range in pupil dilation. When contracted, the pupil is typically pear-shaped, although the bearded seal's is more diagonal. In species that live in shallow water, such as harbor seals and California sea lions, dilation varies little, while the deep-diving elephant seals have much greater variation.

On land, pinnipeds are near-sighted in dim light. This is reduced in bright light, as the retracted pupil reduces the lens and cornea's ability to bend light. They also have a well-developed *tapetum lucidum*, a reflecting layer that increases sensitivity by reflecting light back through the rods. This helps them see in low-light conditions. Ice-living seals like the harp seal have corneas that can tolerate high levels of ultraviolet radiation typical of bright, snowy environments. As such, they do not suffer snow blindness. Pinnipeds appear to have limited color vision, as they lack S-cones. Flexible eye movement has been documented in seals. The extraocular muscles of the walrus are well developed. This and its lack of orbital roof allow it to protrude its eyes and see in both frontal and dorsal directions. Seals release large amounts of mucus to protect their eyes. The corneal epithelium is keratinized and the sclera is thick enough to withstand the pressures of diving. As in many mammals and birds, pinnipeds possess nictitating membranes.

The pinniped ear is adapted for hearing underwater, where it can hear sound frequencies at up to 70,000 Hz. In air, hearing is somewhat reduced in pinnipeds compared to many terrestrial mammals. While they are capable of hearing a wide range of frequencies (e.g. 500 to 32,000 Hz in the northern fur seal, compared to 20 to 20,000 Hz in humans), their airborne hearing sensitivity is weaker overall. One study of three species—the harbor seal, California sea lion and northern elephant seal—found that the sea lion was best adapted for airborne hearing, the harbor seal was equally capable of hearing in air and water, and the elephant seal was better adapted for underwater hearing. Although pinnipeds have a fairly good sense of smell on land, it is useless underwater as their nostrils are closed.

Vibrissae of walrus

Pinnipeds have well-developed tactile senses. Their mystacial vibrissae have ten times the innervation of terrestrial mammals, allowing them to effectively detect vibrations in the water. These vibrations are generated, for example, when a fish swims through water. Detecting vibrations is useful when the animals are foraging and may add to or even replace vision, particularly in darkness. Harbor seals have been observed following varying paths of another seal that swam ahead several minutes before, similar to a dog following a scent trail, and even to discriminate the species and the size of the fish responsible for the trail. Blind ringed seals have even been observed successfully hunting on their own in Lake Saimaa, likely relying on their vibrissae to gain sensory information and catch prey.

Unlike terrestrial mammals, such as rodents, pinnipeds do not move their vibrissae over an object when examining it but instead extend their moveable whiskers and keep them in the same position. By holding their vibrissae steady, pinnipeds are able to maximize their detection ability. The vibrissae of phocids are undulated and wavy while otariid and walrus vibrissae are smooth. Research is ongoing to determine the function, if any, of these shapes on detection ability. The vibrissa's angle relative to the flow, not the shape, however, seems to be the most important factor. The vibrissae of some otariids grow quite long—those of the Antarctic fur seal can reach 41 cm (16 in). Walruses have the most vibrissae, at 600–700 individual hairs. These are important for detecting their prey on the muddy sea floor. In addition to foraging, vibrissae may also play a role in navigation; spotted seals appear to use them to detect breathing holes in the ice.

Diving Adaptations

Diving Weddell seals

Before diving, pinnipeds typically exhale to empty their lungs of half the air and then close their nostrils and throat cartilages to protect the trachea. Their unique lungs have airways that are highly reinforced with cartilaginous rings and smooth muscle, and alveoli that completely deflate during deeper dives. While terrestrial mammals are generally unable to empty their lungs, pinnipeds can reinflate their lungs even after complete respiratory collapse. The middle ear contains sinuses that probably fill with blood during dives, preventing middle ear squeeze. The heart of a seal is moderately flattened to allow the lungs to deflate. The trachea is flexible enough to collapse under pressure. During deep dives, any remaining air in their bodies is stored in the bronchioles and trachea, which prevents them from experiencing decompression sickness, oxygen toxicity and nitrogen narcosis. In addition, seals can tolerate large amounts of lactic acid, which reduces skeletal muscle fatigue during intense physical activity.

The main adaptations of the pinniped circulatory system for diving are the enlargement and increased complexity of veins to increase their capacity. Retia mirabilia form blocks of tissue on the inner wall of the thoracic cavity and the body periphery. These tissue masses, which contain extensive contorted spirals of arteries and thin-walled veins, act as blood reservoirs that increase oxygen stores for use during diving. As with other diving mammals, pinnipeds have high amounts of hemoglobin and myoglobin stored in their blood and muscles. This allows them to stay submerged for long periods of time while still having enough oxygen. Deep-diving species such as elephant seals have blood volumes that make up to 20% of their body weight. When diving, they reduce their heart rate and maintain blood flow only to the heart, brain and lungs. To keep their blood pressure stable, phocids have an elastic aorta that dissipates some energy of each heartbeat.

Thermoregulation

Northern elephant seal resting in water

Pinnipeds conserve heat with their large and compact body size, insulating blubber and fur, and high metabolism. In addition, the blood vessels in their flippers are adapted for countercurrent exchange. Veins containing cool blood from the body extremities surround arteries, which contain warm blood received from the core of the body. Heat from the arterial blood is transferred to the blood vessels, which then recirculate blood back to the core. The same adaptations that conserve heat while in water tend to inhibit heat loss when out of water. To counteract overheating, many species cool off by flipping sand onto their backs, adding a layer of cool, damp sand that enhances heat loss. The northern fur seal pants to help stay cool, while monk seals often dig holes in the sand to expose cooler layers to rest in.

Sleep

Pinnipeds spend many months at a time at sea, so they must sleep in the water. Scientists have recorded them sleeping for minutes at a time while slowly drifting downward in a belly-up orientation. Like other marine mammals, seals sleep in water with half of their brain awake so that they can detect and escape from predators. When they are asleep on land, both sides of their brain go into sleep mode.

Distribution and Habitat

Living pinnipeds mainly inhabit polar and subpolar regions, particularly the North Atlantic, the North Pacific and the Southern Ocean. They are entirely absent from Indo-Malayan waters. Monk seals and some otariids live in tropical and subtropical waters. Seals usually require cool, nutrient-rich waters with temperatures lower than 20 °C (68 °F). Even those that live in warm or tropical climates live in areas that become cold and nutrient rich due to current patterns. Only monk seals live in waters that are not typically cool or rich in nutrients. The Caspian seal and Baikal seal are found in large landlocked bodies of water (the Caspian Sea and Lake Baikal respectively).

Walrus on ice off Alaska. This species has a discontinuous distribution around the Arctic Circle.

As a whole, pinnipeds can be found in a variety of aquatic habitats, including coastal water, open ocean, brackish water and even freshwater lakes and rivers. Most species inhabit coastal areas, though some travel offshore and feed in deep waters off oceanic islands. The Baikal seal is the only freshwater species, though some ringed seals live in freshwater lakes in Russia close to the Baltic sea. In addition, harbor seals may visit estuaries, lakes and rivers and sometimes stay as long as a year. Other species known to enter freshwater include California sea lions and South American sea lions. Pinnipeds also use a number of terrestrial habitats and substrates, both continental and island. In temperate and tropical areas, they haul-out on to sandy and pebble beaches, rocky shores, shoals, mud flats, tide pools and in sea caves. Some species also rest on man-made structures, like piers, jetties, buoys and oil platforms. Pinnipeds may move further inland and rest in sand dunes or vegetation, and may even climb cliffs. Polar-living species haul-out on to both fast ice and pack ice.

Behavior and Life History

Harbor seal hauled out on rock

Pinnipeds have an amphibious lifestyle; they spend most of their lives in the water, but haul-out to mate, raise young, molt, rest, thermoregulate or escape from aquatic predators. Several species are known to migrate vast distances, particularly in response to extreme environmental changes, like El Niño or changes in ice cover. Elephant seals stay at sea 8–10 months a year and migrate between breeding and molting sites. The northern elephant seal has one of the longest recorded migration distance for a mammal, at 18,000–21,000 km (11,000–13,000 mi). Phocids tend to migrate more than otariids. Traveling seals may use various features of their environment to reach their destination including geomagnetic fields, water and wind currents, the position of the sun and moon and the taste and temperature of the water.

Pinnipeds may dive during foraging or to avoid predators. When foraging, Weddell seals typically dive for less than 15 minutes to depths of around 400 m (1,300 ft) but can dive for as long as 73 minutes and to depths of up to 600 m (2,000 ft). Northern elephant seals commonly dive 350–650 m (1,150–2,130 ft) for as long as 20 minutes. They can also dive 1,259–4,100 m (4,131–13,451 ft) and for as long as 62 minutes. The dives of otariids tend to be shorter and less deep. They typically last 5–7 minutes with average depths to 30–45 m (98–148 ft). However, the New Zealand sea lion has been recorded diving to a maximum of 460 m (1,510 ft) and a duration of 12 minutes. Walruses do not often dive very deep, as they feed in shallow water.

Pinnipeds have lifespans averaging 25–30 years. Females usually live longer, as males tend to fight and often die before reaching maturity. The longest recorded lifespans include 43 years for a wild female ringed seal and 46 years for a wild female grey seal. The age at which a pinniped sexually matures can vary from 2–12 years depending on the species. Females typically mature earlier than males.

Foraging and Predation

Steller sea lion with white sturgeon

All pinnipeds are carnivorous and predatory. As a whole, they mostly feed on fish and cephalopods, followed by crustaceans and bivalves, and then zooplankton and endothermic ("warm-blooded") prey like sea birds. While most species are generalist and opportunistic feeders, a few are specialists. Examples include the crabeater seal, which primarily eats krill, the ringed seal, which eats mainly crustaceans, the Ross seal and southern elephant seal, which specialize on squid, and the bearded seal and walrus, which feed on clams and other bottom-dwelling invertebrates. Pinnipeds may hunt solitarily or cooperatively. The former behavior is typical when hunting non-schooling fish, slow-moving or immobile invertebrates or endothermic prey. Solitary foraging species usually exploit coastal waters, bays and rivers. An exception to this is the northern elephant seal, which feeds on fish at great depths in the open ocean. In addition, walruses feed solitarily but are often near other walruses in small or large groups that may surface and dive in unison. When large schools of fish or squid are available, pinnipeds such as certain otariids hunt cooperatively in large groups, locating and herding their prey. Some species, such as California and South American sea lions, may forage with cetaceans and sea birds.

Seals typically consume their prey underwater where it is swallowed whole. Prey that is too large or awkward is taken to the surface to be torn apart. The leopard seal, a prolific predator of penguins, is known to violently swing its prey back and forth until it is decapitated. The elaborately cusped teeth of filter-feeding species, such as crabeater seals, allow them to remove water before they swallow their planktonic food. The walrus is unique in that it consumes its prey by suction feeding, using its tongue to suck the meat of a bivalve out of the shell. While pinnipeds mostly hunt in the water, South American sea lions are known to chase down penguins on land. Some species may swallow stones or pebbles for reasons not understood. Though they can drink seawater, pinnipeds get most of their fluid intake from the food they eat.

Leopard seal capturing emperor penguin

Pinnipeds themselves are subject to predation. Most species are preyed on by the killer whale or orca. To subdue and kill seals, orcas continuously ram them with their heads, slap them with their tails and fling them in the air. They are typically hunted by groups of 10 or fewer whales, but they are occasionally hunted by larger groups or by lone individuals. Pups are more commonly taken by orcas, but adults can be targeted as well. Large sharks are another major predator of pinnipeds—usually the great white shark but also the tiger shark and mako shark. Sharks usually attack by ambushing them from below. The prey usually escapes, and seals are often seen with shark-inflicted wounds. Otariids typically have injuries in the hindquarters, while phocids usually have injuries on the forequarters.

Pinnipeds are also targeted by terrestrial and pagophilic predators. The polar bear is well adapted for hunting Arctic seals and walruses, particularly pups. Bears are known to use sit-and-wait tactics as well as active stalking and pursuit of prey on ice or water. Other terrestrial predators include cougars, brown hyenas and various species of canids, which mostly target the young. Pinnipeds lessen the chance of predation by gathering in groups. Some species are capable of inflicting damaging wounds on their attackers with their sharp canines—an adult walrus is capable of killing polar bears. When out at sea, northern elephant seals dive out of the reach of surface-hunting orcas and white sharks. In the Antarctic, which lacks terrestrial predators, pinniped species spend more time on the ice than their Arctic counterparts. Arctic seals use more breathing holes per individual, appear more restless when hauled out, and rarely defecate on the ice. Ringed seals would build dens underneath fast ice.

Interspecific predation among pinnipeds does occur. The leopard seal is known to prey on numerous other species, especially the crabeater seal. Leopard seals typically target crabeater pups, which form an important part of their diet from November to January. Older crabeater seals commonly bear scars from failed leopard seal attacks; a 1977 study found that 75% of a sample of 85 individual crabeaters had these scars. Walruses, despite being specialized for feeding on bottom-dwelling invertebrates, occasionally prey on Arctic seals. They kill their prey with their long tusks and eat their blubber and skin. Steller sea lions have been recorded eating the pups of harbor seals, northern fur seals and California sea lions. New Zealand sea lions feed on pups of some fur seal species, and the South American sea lion may prey on South American fur seals.

Reproductive Behavior

Northern fur seal breeding colony

The mating system of pinnipeds varies from extreme polygyny to serial monogamy. Of the 33 species, 20 breed on land, and the remaining 13 breed on ice. Species that breed on land are usually polygynous, as females gather in large aggregations and males are able to mate with them as well as defend them from rivals. Polygynous species include elephant seals, grey seals and most otariids. Land-breeding pinnipeds tend to mate on islands where there are fewer terrestrial predators. Few islands are favorable for breeding, and those that are tend to be crowded. Since the land they breed on is fixed, females return to the same sites for many years. The males arrive earlier in the season and wait for them. The males stay on land and try to mate with as many females as they can; some of them will even fast. If a male leaves the beach to feed, he will likely lose mating opportunities and his dominance. Polygynous species also tend to be extremely sexual dimorphic in favor of males. This dimorphism manifests itself in larger chests and necks, longer canines and denser fur—all traits that help males in fights for females. Increased body weight in males increases the length of time they can fast due to the ample energy reserves stored in the blubber. Larger males also likely enjoy access to feeding grounds that smaller ones are unable to access due to their lower thermoregulatory ability and decreased energy stores. In some instances, only the largest males are able to reach the furthest deepest foraging grounds where they enjoy maximum energetic yields that are unavailable to smaller males and females.

Other seals, like the walrus and most phocids, breed on ice with copulation usually taking place in the water (a few land-breeding species also mate in water). Females of these species tend to aggregate less. In addition, since ice is less stable than solid land, breeding sites change location each year, and males are unable to predict where females will stay during the breeding season. Hence polygyny tends to be weaker in ice-breeding species. An exception to this is the walrus, where females form dense aggregations perhaps due to their patchy food sources. Pinnipeds that breed on fast ice tend to cluster together more than those that breed on pack ice. Some of these species are serially monogamous, including the harp seal, crabeater seal and hooded seal. Seals that breed on ice tend to have little or no sexual dimorphism. In lobodontine seals, females are slightly longer than males. Walruses and hooded seals are unique among ice-breeding species in that they have pronounced sexual dimorphism in favor of males.

Male northern elephant seals fighting for dominance and females

Adult male pinnipeds have several strategies to ensure reproductive success. Otariids establish territories containing resources that attract females, such as shade, tide pools or access to water. Territorial boundaries are usually marked by natural breaks in the substrate, and males defend their territorial boundaries with threatening vocalizations and postures, but physical fights are usually avoided. Individuals also return to the same territorial site each breeding season. In certain species, like the Steller sea lion and northern fur seal, a dominant male can maintain a territory for as long as 2–3 months. Females can usually move freely between territories and males are unable to coerce them, but in some species such as the northern fur seal, South American sea lion and Australian sea lion, males can successfully contain females in their territories and prevent them from leaving. In some phocid species, like the harbor seal, Weddell seal and bearded seal, the males have underwater territories called "maritories" near female haul-out areas. These are also maintained by vocalizations. The maritories of Weddell seal males can overlap with female breathing holes in the ice.

Lek systems are known to exist among some populations of walruses. These males cluster around females and try to attract them with elaborate courtship displays and vocalizations. Lekking may also exist among California sea lions, South American fur seals, New Zealand sea lions and harbor seals. In some species, including elephant seals and grey seals, males will try to lay claim to the desired females and defend them from rivals. Elephant seal males establish dominance hierarchies with the highest ranking males—the alpha males—maintaining harems of as many as 30–100 females. These males commonly disrupt the copulations of their subordinates while they themselves can mount without inference. They will, however, break off mating to chase off a rival. Grey seal males usually claim a location among a cluster of females whose members may change over time. Male harp seals, crabeater seals and hooded seals follow and defend lactating females in their vicinity—usually one or two at a time,—and wait for them to reach estrus.

Younger or subdominant male pinnipeds may attempt to achieve reproductive success in other ways. Subadult elephant seals will sneak into female clusters and try to blend in by pulling in their noses. They also harass and attempt to mate with females that head out to the water. In otariid species like the South American and Australian sea lions, non-territorial subadults form "gangs" and cause chaos within the breeding rookeries to increase their chances of mating with females. Alternative mating strategies also exist in young male grey seals, which do have some success. Female pinnipeds do appear to have some choice in mates, particularly in lek-breeding species like the walrus, but also in elephant seals where the males try to dominate all the females that they want to mate with. When a female elephant seal or grey seal is mounted by an unwanted male, she tries to squirm and get away, while croaking and slapping him with her tail. This commotion attracts other males to the scene, and the most dominant will end the copulation and attempt to mate with the female himself. Dominant female elephant seals stay in the center of the colony where they are more likely to mate with a dominant male, while peripheral females are more likely to mate with subordinates. Female Steller sea lions are known to solicit mating with their territorial males.

Birth and Parenting

With the exception of the walrus, which has five- to six-year-long inter-birth intervals, female pinnipeds enter estrous shortly after they give birth. All species go through delayed implantation, wherein the embryo remains in suspended development for weeks or months before it is implanted in the uterus. Delayed implantation postpones the birth of young until the female hauls-out on land or until conditions for birthing are favorable. Gestation in seals (including delayed implantation) typically lasts a year. For most species, birthing takes place in the spring and summer months. Typically, single pups are born; twins are uncommon and have high mortality rates. Pups of most species are born precocial. Unlike terrestrial mammals, pinniped milk has little to no lactose.

Mother pinnipeds have different strategies for maternal care and lactation. Phocids such as elephant seals, grey seals and hooded seals remain on land or ice and fast during their relatively short lactation period—four days for the hooded seal and five weeks for elephant seals. The milk of these species consist of up to 60% fat, allowing the young to grow fairly quickly. In particular, northern elephant seal pups gain 4 kg (9 lb) each day before they are weaned. Some pups may try to steal extra milk from other nursing mothers and gain weight more quickly than others. Alloparenting occurs in these fasting species; while most northern elephant seal mothers nurse their own pups and reject nursings from alien pups, some do accept alien pups with their own.

For otariids and some phocids like the harbor seal, mothers fast and nurse their pups for a few days at a time. In between nursing bouts, the females leave their young onshore to forage at sea. These foraging trips may last anywhere between a day and two weeks, depending on the abundance of food and the distance of foraging sites. While their mothers are away, the pups will fast. Lactation in otariids may last 6–11 months; in the Galápagos fur seal it can last as long as 3 years. Pups of these species are weaned at lower weights than their phocid counterparts. Walruses are unique in that mothers nurse their young at sea. The female rests at the surface with its head held up, and the young suckle upside down.

Male pinnipeds generally play little role in raising the young. Male walruses may help inexperienced young as they learn to swim, and have even been recorded caring for orphans. Male California sea lions have been observed to help shield swimming pups from predators. Males can also pose threats to the safety of pups. In terrestrially breeding species, pups may get crushed by fighting males. Subadult male South America sea lions sometimes abduct pups from their mothers and treat them like adult males treat females. This helps them gain experience in controlling females. Pups can get severely injured or killed during abductions.

Communication

Walrus males are known to use vocalizations to attract mates.

Pinnipeds can produce a number of vocalizations such as barks, grunts, rasps, rattles, growls, creaks, warbles, trills, chirps, chugs, clicks and whistles. Vocals are produced both in air and underwater. Otariids are more vocal on land, while phocids are more vocal in water. Antarctic seals are more vocal on land or ice than Arctic seals due to a lack of terrestrial and pagophliic predators like the polar bear. Male vocals are usually of lower frequencies than those of the females. Vocalizations are particularly important during the breeding seasons. Dominant male elephant seals advertise their status and threaten rivals with "clap-threats" and loud drum-like calls that may be modified by the proboscis. Male otariids have strong barks, growls, roars and "whickers". Male walruses are known to produce distinctive gong-like calls when attempting to attract females. They can also create somewhat musical sounds with their inflated throats.

The Weddell seal has perhaps the most elaborate vocal repertoire with separate sounds for airborne and underwater contexts. Underwater vocals include trills, chugs, chirps, chugs and knocks. The calls appear to contain prefixes and suffixes that serve to emphasize a message. The underwater vocals of Weddell seals can last 70 seconds, which is long for a marine mammal call. Some calls have around seven rhythm patterns and are comparable to birdsongs and whalesongs. Similar calls have been recorded in other lobodontine seals and in bearded seals. In some pinniped species, there appear to be geographic differences in vocalizations, known as dialects, while certain species may even have individual variations in expression. These differences are likely important for mothers and pups who need to remain in contact on crowded beaches. Otariid females and their young use mother-pup attraction calls to help them reunite when the mother returns from foraging at sea. Female pinnipeds are also known to bellow when protecting their young.

While most vocals are audible to the human ear, a captive leopard seal was recorded making ultrasonic calls underwater. In addition, the vocals of northern elephant seals may produce infrasonic vibrations. Non-vocal communication is not as common in pinnipeds as in cetaceans. Nevertheless, when disturbed by intruders harbor seals and Baikal seals may slap their fore-flippers against their bodies as warnings. Teeth chattering, hisses and exhalations are also made as aggressive warnings. Visual displays also occur: Weddell seals will make an S-shaped posture when patrolling under the ice, and Ross seals will display the stripes on their chests and teeth when approached. Male hooded seals use their inflatable nasal membranes to display to and attract females.

Intelligence

In a match-to-sample task study, a single California sea lion was able to demonstrate an understanding of symmetry, transitivity and equivalence; a second seal was unable to complete the tasks. They demonstrate the ability to understand simple syntax and commands when taught an artificial sign language, though they only rarely used the signs semantically or logically. In 2011, a captive California sea lion named Ronan was recorded bobbing its head in synchrony to musical rhythms. This "rhythmic entrainment" was previously seen only in humans, parrots and other birds possessing vocal mimicry. In 1971, a captive harbor seal named Hoover was trained to imitate human words, phrases and laughter. For sea lions used in entertainment, trainers toss a ball at the animal so it may accidentally balance it or hold the ball on its nose, thereby gaining an understanding of the behavior desired. It may require a year to train a sea lion to perform a trick for the public. Its long-term memory allows it to perform a trick after at least three months of non-performance.

Human Relations

Cultural Depictions

Engraving of the "The Walrus and the Carpenter"

Various human cultures have depicted pinnipeds for millennia. The anthropologist, A. Asbjørn Jøn, has analysed beliefs of the Celts of Orkney and Hebrides who believed in the selkie—seals that could change into humans and walk on land. Seals are also of great importance in the culture of the Inuit. In Inuit mythology, the goddess Sedna rules over the sea and marine animals. She is depicted as a mermaid, occasionally with a seal's lower body. In one legend, seals, whales and other marine mammals were formed from her severed fingers. One of the earliest Ancient Greek coins depicts the head of a seal, and the animals were mentioned by Homer and Aristotle. The Greeks believed that seals loved both the sea and sun and were considered to be under the protection of the gods Poseidon and Apollo. The Moche people of ancient Peru worshipped the sea and its animals, and often depicted sea lions in their art. In modern popular culture, pinnipeds are often depicted as comical figures based on their performances in zoos, circuses and marine mammal parks.

In captivity

Captive sea lion at Kobe Oji Zoo, Japan

Pinnipeds have been kept in captivity since at least the 17th century and can be found in facilities around the world. Their large size and playfulness make them popular attractions. Some exhibits have rocky backgrounds with artificial haul-out sites and a pool, while others have pens with small rocky, elevated shelters where the animals can dive into their pools. More elaborate exhibits contain deep pools that can be viewed underwater with rock-mimicking cement as haul-out areas. The most common pinniped species kept in captivity is the California sea lion as it is abundant and easy to train. These animals are used to perform tricks and entertain visitors. Other species popularly kept in captivity include the grey seal and harbor seal. Larger animals like walruses and Steller sea lions are much less common. Some organizations, such as the Humane Society of the United States and World Animal Protection, object to keeping pinnipeds and other marine mammals in captivity. They state that the exhibits could not be large enough to house animals that have evolved to be migratory, and a pool could never replace the size and biodiversity of the ocean. They also oppose using sea lions for entertainment, claiming the tricks performed are "exaggerated variations of their natural behaviors" and distract the audience from the animal's unnatural environment.

California sea lions are used in military applications by the U.S. Navy Marine Mammal Program, including detecting naval mines and enemy divers. In the Persian Gulf, the animals can swim behind divers approaching a U.S. naval ship and attach a clamp with a rope to the diver's leg. Navy officials say that the sea lions can do this in seconds, before the enemy realizes what happened. Organizations like PETA believe that such operations put the animals in danger. The Navy insists that the sea lions are removed once their mission is complete.

Conservation and Management Issues

As of 2013, International Union for Conservation of Nature (IUCN) recognizes 35 pinniped species. Two species—the Japanese sea lion and the Caribbean monk seal—are recently extinct, and ten more are considered at risk, as they are ranked "Critically Endangered" (the Mediterranean and Hawaiian monk seals), "Endangered" (Galápagos fur seal, Australian sea lion, Caspian seal and Galápagos sea lion), and "Vulnerable" (northern fur seal, hooded seal and New Zealand sea lion). Three species—the walrus, the ribbon seal, and the spotted seal—have a "Data Deficient" ranking. Species that live in polar habitats are vulnerable to the effects of recent and ongoing climate change, particularly declines in sea ice. There has been some debate over the cause of the decline of Steller sea lions in Alaska since the 1970s.

Humans have hunted seals since the Stone Age. Originally, seals were hit with clubs during haul-out. Eventually, seal hunters used harpoons to spear the animals from boats out at sea, and hooks for killing pups on ice or land. They were also trapped in nets. The use of firearms in seal hunting during the modern era drastically increased the number of killings. Pinnipeds are typically hunted for their meat and blubber. The skins of fur seals and phocids are made into coats, and the tusks of walruses continue to be used for carvings or as ornaments. There is a distinction between the subsistence hunting of seals by indigenous peoples of the Arctic and commercial hunting: subsistence hunters typically use seal products for themselves and depend on them for survival. National and international authorities have given special treatment to aboriginal hunters since their methods of killing are seen as less destructive and wasteful. This distinction is being questioned as indigenous people are using more modern weaponry and mechanized transport to hunt with, and are selling seal products in the marketplace. Some anthropologists argue that the term "subsistence" should also apply to these cash-based exchanges as long as they take place within local production and consumption. More than 100,000 phocids (especially ringed seals) as well as around 10,000 walruses are harvested annually by native hunters.

Protests of Canada's seal hunts

Commercial sealing was historically just as important an industry as whaling. Exploited species included harp seals, hooded seals, Caspian seals, elephant seals, walruses and all species of fur seal. The scale of seal harvesting decreased substantially after the 1960s, after the Canadian government reduced the length of the hunting season and implemented measures to protect adult females. Several species that were commercially exploited have rebounded in numbers; for example, Antarctic fur seals may be as numerous as they were prior to harvesting. The northern elephant seal was hunted to near extinction in the late 19th century, with only a small population remaining on Guadalupe Island. It has since recolonized much of its historic range, but has a population bottleneck. Conversely, the Mediterranean monk seal was extirpated from much of its former range, which stretched from the Mediterranean to the Black Sea and northwest Africa, and only remains in the northeastern Mediterranean and some parts of northwest Africa.

Several species of pinniped continue to be harvested. The Convention for the Conservation of Antarctic Seals allows limited hunting of crabeater seals, leopard seals and Weddell seals. However, Weddell seal hunting is prohibited between September and February if the animal is over one year of age, to ensure breeding stocks are healthy. Other species protected are southern elephant seals, Ross seals and Antarctic fur seals. The Government of Canada permits the hunting of harp seals. This has been met with controversy and debate. Proponents of seal hunts insist that the animals are killed humanely and the white-coated pups are not taken, while opponents argue that it is irresponsible to kill harp seals as they are already threatened by declining habitat.

The Caribbean monk seal has been killed and exploited by Europeans settlers and their descendants since 1494, starting with Christopher Columbus himself. The seals were easy targets for organized sealers, fishermen, turtle hunters and buccaneers because they evolved with little pressure from terrestrial predators and were thus "genetically tame". In the Bahamas, as many as 100 seals were slaughtered in one night. In the mid-nineteenth century, the species was thought to have gone extinct until a small colony was found near the Yucatán Peninsula in 1886. Seal killings continued, and the last reliable report of the animal alive was in 1952. The IUCN declared it extinct in 1996. The Japanese sea lion was common around the Japanese islands, but overexploitation and competition from fisheries drastically decreased the population in the 1930s. The last recorded individual was a juvenile in 1974.

Grey seal on beach occupied by humans. Pinnipeds and humans may compete for space and resources.

Some species have become so numerous that they conflict with local people. In the United States, pinnipeds and other marine mammals are protected under the Marine Mammal Protection Act of

1972 (MMPA). Since that year, California sea lion populations have risen to 250,000. These animals began exploiting more man-made environments, like docks, for haul-out sites. Many docks are not designed to withstand the weight of several resting sea lions, which causes major tilting and other problems. Wildlife managers have used various methods to control the animals, and some city officials have redesigned docks so they can better withstand them. Sea lions also conflict with fisherman since both depend on the same fish stocks. In 2007, MMPA was amended to permit the lethal removal of sea lions from salmon runs at Bonneville Dam. The 2007 law seeks to relieve pressure on the crashing Pacific Northwest salmon populations. Wildlife officials have unsuccessfully attempted to ward off the sea lions using bombs, rubber bullets and bean bags. Efforts to chase sea lions away from the area have also proven ineffective. Critics like the Humane Society object to the killing of the sea lions, claiming that hydroelectric dams pose a greater threat to the salmon. Similar conflicts have existed in South Africa with brown fur seals. In the 1980s and 1990s, South African politicians and fisherman demanded that the fur seals be culled, believing that the animals competed with commercial fisheries. Scientific studies found that culling fur seals would actually have a negative effect on the fishing industry, and the culling option was dropped in 1993.

Pinnipeds can also be threatened by humans in more indirect ways. They are unintentionally caught in fishing nets by commercial fisheries and accidentally swallow fishing hooks. Gillnetting and Seine netting is a significant cause of mortality in seals and other marine mammals. Species commonly entangled include California sea lions, Hawaiian monk seals, northern fur seals and brown fur seals. Pinnipeds are also affected by marine pollution. High levels of organic chemicals accumulate in these animals since they are near the top of food chains and have large reserves of blubber. Lactating mothers can pass the toxins on to their young. These pollutants can cause gastrointestinal cancers, decreased reproductivity and greater vulnerability to infectious diseases. Other man-made threats include habitat destruction by oil and gas exploitation, encroachment by boats and underwater noise.

Sea Otter

The sea otter (*Enhydra lutris*) is a marine mammal native to the coasts of the northern and eastern North Pacific Ocean. Adult sea otters typically weigh between 14 and 45 kg (31 and 99 lb), making them the heaviest members of the weasel family, but among the smallest marine mammals. Unlike most marine mammals, the sea otter's primary form of insulation is an exceptionally thick coat of fur, the densest in the animal kingdom. Although it can walk on land, the sea otter lives mostly in the ocean.

The sea otter inhabits offshore environments, where it dives to the sea floor to forage. It preys mostly on marine invertebrates such as sea urchins, various molluscs and crustaceans, and some species of fish. Its foraging and eating habits are noteworthy in several respects. First, its use of rocks to dislodge prey and to open shells makes it one of the few mammal species to use tools. In most of its range, it is a keystone species, controlling sea urchin populations which would otherwise inflict extensive damage to kelp forest ecosystems. Its diet includes prey species that are also valued by humans as food, leading to conflicts between sea otters and fisheries.

Sea otters, whose numbers were once estimated at 150,000–300,000, were hunted extensively for their fur between 1741 and 1911, and the world population fell to 1,000–2,000 individuals living in a fraction of their historic range. A subsequent international ban on hunting, conservation efforts, and reintroduction programs into previously populated areas have contributed to numbers rebounding, and the species now occupies about two-thirds of its former range. The recovery of the sea otter is considered an important success in marine conservation, although populations in the Aleutian Islands and California have recently declined or have plateaued at depressed levels. For these reasons, the sea otter remains classified as an endangered species.

Taxonomy

Mother sea otter with rare twin pups, Morro Bay, California. Sea otter twin births are rare, and the high demands on the mother usually result in one pup being abandoned.

Sea otter ln Morro Bay, California

The first scientific description of the sea otter is contained in the field notes of Georg Steller from 1751, and the species was described by Linnaeus in his *Systema Naturae* of 1758. Originally named *Lutra marina*, it underwent numerous name changes before being accepted as *Enhydra lutris* in 1922.

The sea otter was formerly sometimes referred to as the "sea beaver", being the marine fur-bearer similar in commercial value to the terrestrial beaver. Rodents (of which the beaver is one) are not closely related to otters, which are carnivores. It is not to be confused with the marine otter, a rare otter species native to the southern west coast of South America. A number of other otter species, while predominantly living in fresh water, are commonly found in marine coastal habitats. The extinct sea mink of northeast North America is another mustelid that had adapted to a marine environment.

Evolution

The sea otter is the heaviest (the giant otter is longer, but significantly slimmer) member of the family Mustelidae, a diverse group that includes the 13 otter species and terrestrial animals such as weasels, badgers, and minks. It is unique among the mustelids in not making dens or burrows, in having no functional anal scent glands, and in being able to live its entire life without leaving the water. The only member of the genus *Enhydra*, the sea otter is so different from other mustelid species that, as recently as 1982, some scientists believed it was more closely related to the earless seals. Genetic analysis indicates the sea otter and its closest extant relatives, which include the African speckle-throated otter, European otter, African clawless otter and oriental small-clawed otter, shared an ancestor approximately 5 million years ago (Mya).

Fossil evidence indicates the *Enhydra* lineage became isolated in the North Pacific approximately 2 Mya, giving rise to the now-extinct *Enhydra macrodonta* and the modern sea otter, *Enhydra lutris*. The sea otter evolved initially in northern Hokkaidō and Russia, and then spread east to the Aleutian Islands, mainland Alaska, and down the North American coast. In comparison to cetaceans, sirenians, and pinnipeds, which entered the water approximately 50, 40, and 20 Mya, respectively, the sea otter is a relative newcomer to a marine existence. In some respects, though, the sea otter is more fully adapted to water than pinnipeds, which must haul out on land or ice to give birth.

One related species has been described, *Enhydra reevei*, from the Pleistocene of East Anglia. The holotype, a lower carnassial, was in the Norwich Castle Museum but seems to be lost. Only one more specimen, an extremely worn lower carnassial, is known.

Subspecies

The three recognized subspecies, which vary in body size and in some skull and dental characteristics, are:

Subspecies	Trinomial authority	Common names	Description	Range	Synonyms
E. l. lutris (Nominate subspecies)	Linnaeus, 1758	**Common sea otter** **Asian sea otter** **Commander sea otter** **Kuril sea otter**	The largest subspecies, with a wide skull and short nasal bones	Kuril Islands to the Commander Islands in the western Pacific Ocean	*gracilis* (Bechstein, 1800) *kamtschatica* (Dybowski, 1922) *marina* (Erxleben, 1777) *orientalis* (Oken, 1816) *stelleri* (Lesson, 1827)

E. l. nereis	Merriam, 1904	**Southern sea otter** **California sea otter**	Has a narrower skull with a long rostrum and small teeth	Coast of central California	
E. l. kenyoni	Wilson, 1991	**Northern sea otter**		Alaska and the Pacific west coast from the Aleutian islands to British Columbia, Washington, and northern Oregon, after being extirpated from southern British Columbia due to overhunting, it has since been reintroduced off Vancouver Island and the Olympic Peninsula.	

The reintroduction effort off the Oregon coast was not successful. However, reintroductions in 1969 and 1970 off the Washington coast were very successful and sea otters have been expanding their range since. They have now entered the Strait of Juan de Fuca and can be found almost as far east as Pillar Point. Individuals have even been seen in the San Juan Islands and northern Puget Sound.

Physical Characteristics

Sea otter's thick fur makes its body appear much plumper on land than in the water

The sea otter is one of the smallest marine mammal species, but it is the heaviest mustelid. Male sea otters usually weigh 22 to 45 kg (49 to 99 lb) and are 1.2 to 1.5 m (3 ft 11 in to 4 ft 11 in) in length, though specimens to 54 kg (119 lb) have been recorded. Females are smaller, weighing 14 to 33 kg (31 to 73 lb) and measuring 1.0 to 1.4 m (3 ft 3 in to 4 ft 7 in) in length. For its size, the male otter's baculum is very large, massive and bent upwards, measuring 150 mm (5.9 in) in length and 15 mm (0.59 in) at the base.

Unlike most other marine mammals, the sea otter has no blubber and relies on its exceptionally thick fur to keep warm. With up to 150,000 strands of hair per square centimeter (nearly one million per sq in), its fur is the densest of any animal. The fur consists of long, waterproof guard hairs and short underfur; the guard hairs keep the dense underfur layer dry. Cold water is kept completely away from the skin and heat loss is limited. The fur is thick year-round, as it is shed and replaced gradually rather than in a distinct molting season. As the ability of the guard hairs to repel water depends on utmost cleanliness, the sea otter has the ability to reach and groom the fur on any part of its body, taking advantage of its loose skin and an unusually supple skeleton. The coloration of the pelage is usually deep brown with silver-gray speckles, but it can range from yellowish or grayish brown to almost black. In adults, the head, throat, and chest are lighter in color than the rest of the body.

The sea otter displays numerous adaptations to its marine environment. The nostrils and small ears can close. The hind feet, which provide most of its propulsion in swimming, are long, broadly flattened, and fully webbed. The fifth digit on each hind foot is longest, facilitating swimming while on its back, but making walking difficult. The tail is fairly short, thick, slightly flattened, and muscular. The front paws are short with retractable claws, with tough pads on the palms that enable gripping slippery prey. The bones show osteosclerosis, increasing their density to reduce buoyancy.

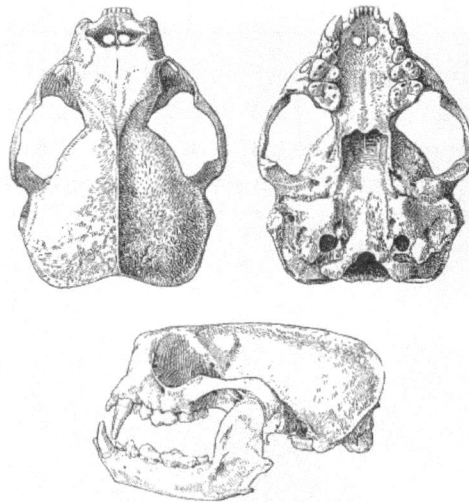

Skull, as illustrated by N. N. Kondakov

The sea otter propels itself underwater by moving the rear end of its body, including its tail and hind feet, up and down, and is capable of speeds of up to 9 km/h (5.6 mph). When underwater, its body is long and streamlined, with the short forelimbs pressed closely against the chest. When at the surface, it usually floats on its back and moves by sculling its feet and tail from side to side. At rest, all four limbs can be folded onto the torso to conserve heat, whereas on particularly hot days, the hind feet may be held underwater for cooling. The sea otter's body is highly buoyant because of its large lung capacity – about 2.5 times greater than that of similar-sized land mammals – and the air trapped in its fur. The sea otter walks with a clumsy, rolling gait on land, and can run in a bounding motion.

Long, highly sensitive whiskers and front paws help the sea otter find prey by touch when waters are dark or murky. Researchers have noted when they approach in plain view, sea otters react more rapidly when the wind is blowing towards the animals, indicating the sense of smell is more important than sight as a warning sense. Other observations indicate the sea otter's sense of sight is useful above and below the water, although not as good as that of seals. Its hearing is neither particularly acute nor poor.

An adult's 32 teeth, particularly the molars, are flattened and rounded, designed to crush rather than cut food. Seals and sea otters are the only carnivores with two pairs of lower incisor teeth rather than three; the adult dental formula is 3.1.3.12.1.3.2

The sea otter has a metabolic rate two or three times that of comparatively sized terrestrial mammals. It must eat an estimated 25 to 38% of its own body weight in food each day to burn the calories necessary to counteract the loss of heat due to the cold water environment. Its digestive efficiency is estimated at 80 to 85%, and food is digested and passed in as little as three hours. Most of its need for water is met through food, although, in contrast to most other marine mammals, it also drinks seawater. Its relatively large kidneys enable it to derive fresh water from sea water and excrete concentrated urine.

A sea otter has two types of fur, the underfur and the guard hair. The shape of these different hair fibers connect to trap air between them. This allows them to maintain their body heat without the blubber other sea mammals use.

Behavior

The sea otter is diurnal. It has a period of foraging and eating in the morning, starting about an hour before sunrise, then rests or sleeps in mid-day. Foraging resumes for a few hours in the afternoon and subsides before sunset, and a third foraging period may occur around midnight. Females with pups appear to be more inclined to feed at night. Observations of the amount of time a sea otter must spend each day foraging range from 24 to 60%, apparently depending on the availability of food in the area.

Sea otters spend much of their time grooming, which consists of cleaning the fur, untangling knots, removing loose fur, rubbing the fur to squeeze out water and introduce air, and blowing air into the fur. To casual observers, it appears as if the animals are scratching, but they are not known to have lice or other parasites in the fur. When eating, sea otters roll in the water frequently, apparently to wash food scraps from their fur.

Foraging

The sea otter hunts in short dives, often to the sea floor. Although it can hold its breath for up to five minutes, its dives typically last about one minute and no more than four. It is the only marine animal capable of lifting and turning over rocks, which it often does with its front paws when searching for prey. The sea otter may also pluck snails and other organisms from kelp and dig deep into underwater mud for clams. It is the only marine mammal that catches fish with its forepaws rather than with its teeth.

Under each foreleg, the sea otter has a loose pouch of skin that extends across the chest. In this pouch (preferentially the left one), the animal stores collected food to bring to the surface. This pouch also holds a rock, unique to the otter, that is used to break open shellfish and clams. There, the sea otter eats while floating on its back, using its forepaws to tear food apart and bring it to its mouth. It can chew and swallow small mussels with their shells, whereas large mussel shells may be twisted apart. It uses its lower incisor teeth to access the meat in shellfish. To eat large sea urchins, which are mostly covered with spines, the sea otter bites through the underside where the spines are shortest, and licks the soft contents out of the urchin's shell.

The sea otter's use of rocks when hunting and feeding makes it one of the few mammal species to use tools. To open hard shells, it may pound its prey with both paws against a rock on its chest. To pry an abalone off its rock, it hammers the abalone shell using a large stone, with observed rates of 45 blows in 15 seconds. Releasing an abalone, which can cling to rock with a force equal to 4,000 times its own body weight, requires multiple dives.

Social Structure

Sleeping sea otters holding paws, photographed at the Vancouver Aquarium, Vancouver, British Columbia, Canada. Note the high buoyancy of the animals' bodies.

Although each adult and independent juvenile forages alone, sea otters tend to rest together in single-sex groups called rafts. A raft typically contains 10 to 100 animals, with male rafts being larger than female ones. The largest raft ever seen contained over 2000 sea otters. To keep from drifting out to sea when resting and eating, sea otters may wrap themselves in kelp.

A male sea otter is most likely to mate if he maintains a breeding territory in an area that is also favored by females. As autumn is the peak breeding season in most areas, males typically defend their territory only from spring to autumn. During this time, males patrol the boundaries of their territories to exclude other males, although actual fighting is rare. Adult females move freely between male territories, where they outnumber adult males by an average of five to one. Males that do not have territories tend to congregate in large, male-only groups, and swim through female areas when searching for a mate.

The species exhibits a variety of vocal behaviors. The cry of a pup is often compared to that of a seagull. Females coo when they are apparently content; males may grunt instead. Distressed or frightened adults may whistle, hiss, or in extreme circumstances, scream.

Although sea otters can be playful and sociable, they are not considered to be truly social animals. They spend much time alone, and each adult can meet its own needs in terms of hunting, grooming, and defense.

Reproduction and Lifecycle

During mating, the male bites the nose of the female, often bloodying and scarring it

Sea otters are polygynous: males have multiple female partners. However, temporary pair-bonding occurs for a few days between a female in estrus and her mate. Mating takes place in the water and can be rough, the male biting the female on the muzzle – which often leaves scars on the nose – and sometimes holding her head under water.

Births occur year-round, with peaks between May and June in northern populations and between January and March in southern populations. Gestation appears to vary from four to twelve months, as the species is capable of delayed implantation followed by four months of pregnancy. In California, sea otters usually breed every year, about twice as often as those in Alaska.

Birth usually takes place in the water and typically produces a single pup weighing 1.4 to 2.3 kg (3 to 5 lb). Twins occur in 2% of births; however, usually only one pup survives. At birth, the eyes are open, ten teeth are visible, and the pup has a thick coat of baby fur. Mothers have been observed to lick and fluff a newborn for hours; after grooming, the pup's fur retains so much air, the pup floats like a cork and cannot dive. The fluffy baby fur is replaced by adult fur after about 13 weeks.

A mother floats with her pup on her chest. Georg Steller wrote, "They embrace their young with an affection that is scarcely credible."

Nursing lasts six to eight months in Californian populations and four to twelve months in Alaska, with the mother beginning to offer bits of prey at one to two months. The milk from a sea otter's two abdominal nipples is rich in fat and more similar to the milk of other marine mammals than to that of other mustelids. A pup, with guidance from its mother, practices swimming and diving for several weeks before it is able to reach the sea floor. Initially, the objects it retrieves are of little food value, such as brightly colored starfish and pebbles. Juveniles are typically independent at six to eight months, but a mother may be forced to abandon a pup if she cannot find enough food for it; at the other extreme, a pup may nurse until it is almost adult size. Pup mortality is high, particularly during an individual's first winter – by one estimate, only 25% of pups survive their first year. Pups born to experienced mothers have the highest survival rates.

Females perform all tasks of feeding and raising offspring, and have occasionally been observed caring for orphaned pups. Much has been written about the level of devotion of sea otter mothers for their pups – a mother gives her infant almost constant attention, cradling it on her chest away from the cold water and attentively grooming its fur. When foraging, she leaves her pup floating on the water, sometimes wrapped in kelp to keep it from floating away; if the pup is not sleeping, it cries loudly until she returns. Mothers have been known to carry their pups for days after the pups' deaths.

Females become sexually mature at around three or four years of age and males at around five; however, males often do not successfully breed until a few years later. A captive male sired offspring at age 19. In the wild, sea otters live to a maximum age of 23 years, with average lifespans of 10–15 years for males and 15–20 years for females. Several captive individuals have lived past 20 years, and a female at the Seattle Aquarium died at the age of 28 years. Sea otters in the wild often develop worn teeth, which may account for their apparently shorter lifespans.

There are several documented cases in which male sea otters have forcibly copulated with juvenile harbor seals, sometimes resulting in death. Similarly, forced copulation by Sea Otters involving animals other than Pacific Harbor Seals has occasionally been reported.

Population and Distribution

Sea otter floating in Morro Bay, California

Sea otters live in coastal waters 15 to 23 meters (50 to 75 ft) deep, and usually stay within a kilometer (mi) of the shore. They are found most often in areas with protection from the most severe ocean winds, such as rocky coastlines, thick kelp forests, and barrier reefs. Although they are most strongly associated with rocky substrates, sea otters can also live in areas where the sea floor consists primarily of mud, sand, or silt. Their northern range is limited by ice, as sea otters can survive amidst drift ice but not land-fast ice. Individuals generally occupy a home range a few kilometers long, and remain there year-round.

The sea otter population is thought to have once been 150,000 to 300,000, stretching in an arc across the North Pacific from northern Japan to the central Baja California Peninsula in Mexico. The fur trade that began in the 1740s reduced the sea otter's numbers to an estimated 1,000 to 2,000 members in 13 colonies. In about two-thirds of its former range, the species is at varying levels of recovery, with high population densities in some areas and threatened populations in others. Sea otters currently have stable populations in parts of the Russian east coast, Alaska, British Columbia, Washington, and California, with reports of recolonizations in Mexico and Japan. Population estimates made between 2004 and 2007 give a worldwide total of approximately 107,000 sea otters.

Russia

Currently, the most stable and secure part of the sea otter's range is Russia. Before the 19th century, around 20,000 to 25,000 sea otters lived near the Kuril Islands, with more near Kamchatka and the Commander Islands. After the years of the Great Hunt, the population in these areas, currently part of Russia, was only 750. By 2004, sea otters had repopulated all of their former habitat in these areas, with an estimated total population of about 27,000. Of these, about 19,000 are at the Kurils, 2,000 to 3,500 at Kamchatka and another 5,000 to 5,500 at the Commander Islands. Growth has slowed slightly, suggesting the numbers are reaching carrying capacity.

Sea otter, Kenai Fjords National Park, Alaska

Alaska

Alaska is the heartland of the sea otter's range. In 1973, the population in Alaska was estimated at between 100,000 and 125,000 animals. By 2006, though, the Alaska population had fallen to an estimated 73,000 animals. A massive decline in sea otter populations in the Aleutian Islands accounts for most of the change; the cause of this decline is not known, although orca predation is suspected. The sea otter population in Prince William Sound was also hit hard by the Exxon Valdez oil spill, which killed thousands of sea otters in 1989.

British Columbia

Along the North American coast south of Alaska, the sea otter's range is discontinuous. A remnant population survived off Vancouver Island into the 20th century, but it died out despite the 1911 international protection treaty, with the last sea otter taken near Kyuquot in 1929. From 1969 to 1972, 89 sea otters were flown or shipped from Alaska to the west coast of Vancouver Island.

John Webber's *Sea Otter, circa* 1788

This population expanded to over 3,200 in 2004, and their range on the island's west coast expanded from Cape Scott in the north to Barkley Sound to the south. In 1989, a separate colony was discovered in the central British Columbia coast. It is not known if this colony, which numbered about 300 animals in 2004, was founded by transplanted otters or by survivors of the fur trade.

The status of the sea otters has improved since 2004 with a report of 4,700 in 2008 that improved their status to "special concern" in Canada. They currently occupy much of the exposed west coast of Vancouver Island and parts of the central mainland BC coast.

Washington

In 1969 and 1970, 59 sea otters were translocated from Amchitka Island to Washington. Annual surveys between 2000 and 2004 have recorded between 504 and 743 individuals, and their range is in the Olympic Peninsula from just south of Destruction Island to Pillar Point. In Washington, sea otters are found almost exclusively on the outer coasts. They can swim as close as six feet off shore along the Olympic coast. Reported sightings of sea otters in the San Juan Islands and Puget Sound almost always turn out to be North American river otters, which are commonly seen along the seashore. However, biologists have confirmed isolated sightings of sea otters in these areas since the mid-1990s.

California

The historic population of California sea otters was estimated at 16,000 before the fur trade began. California's sea otters are the descendants of a single colony of about 50 southern sea otters discovered near Bixby Bridge in Big Sur in 1938. Their principal range has gradually expanded and extends from Pigeon Point in San Mateo County to Santa Barbara County.

Sea otter nursing pup. California has almost 3,000 sea otters, descendants of about 50 individuals discovered in 1938.

For southern sea otters to be considered for removal from threatened species listing, the population would have to exceed 3,090 for three consecutive years. The most recent (spring 2014) United States Geological Survey (USGS) California sea otter survey count is 2,944, almost flat for the last five years. There has been some contraction from the northern (now Pigeon Point) and southern limits of the sea otter's range apparently related to lethal shark bites. The 2013 USGS survey found 2,941 California sea otters, a slight increase from 2012 but a portion of the increase is artificial because the count included, for the first time, the SNI population which had recovered to 59 individuals. The California sea otter census in 2012 was 2,792, down from the peak spring 2007 census of 3,026 sea otters, but up from the recent low of 2,711 in 2010.

In the late 1980s, the U.S. Fish and Wildlife Service (USFWS) relocated about 140 Californian sea otters to San Nicolas Island (SNI) in southern California, in the hope of establishing a reserve population should the mainland be struck by an oil spill. To the surprise of biologists, the San Nicolas sea otters mostly swam back to the mainland. By 2005, only 30 sea otters remained at San Nicolas, although they were slowly increasing as they thrived on the abundant prey around the island. The plan that authorized the translocation program had predicted the carrying capacity would be reached within five to 10 years. The spring 2014 SNI count was 68 sea otters, continuing a 5-year positive trend of over 16% per year.

When the FWS implemented the translocation program, it also attempted to implement "zonal management" of the Californian population. To manage the competition between sea otters and fisheries, it declared an "otter-free zone" stretching from Point Conception to the Mexican border. In this zone, only San Nicolas Island was designated as sea otter habitat, and sea otters found elsewhere in the area were supposed to be captured and relocated. These plans were abandoned after many translocated otters died and also as it proved impractical to capture the hundreds of otters which ignored regulations and swam into the zone. However, after engaging in a period of public commentary in 2005, the Fish and Wildlife Service failed to release a formal decision on the issue. Then, in response to lawsuits filed by lthe Santa Barbara-based Environmental Defense Center and the Otter Project, on December 19, 2012 the USFWS declared that the "no otter zone" experiment was a failure, and will protect the otters re-colonizing the coast south of Point Conception as threatened species.

Sea otters were once numerous in San Francisco Bay. Historical records revealed the Russian-American Company sneaked Aleuts into San Francisco Bay multiple times, despite the Spanish capturing or shooting them while hunting sea otters in the estuaries of San Jose, San Mateo,

San Bruno and around Angel Island. The founder of Fort Ross, Ivan Kuskov, finding otters scarce on his second voyage to Bodega Bay in 1812, sent a party of Aleuts to San Francisco Bay, where they met another Russian party and an American party, and caught 1,160 sea otters in three months. By 1817, sea otters in the area were practically eliminated and the Russians sought permission from the Spanish and the Mexican governments to hunt further and further south of San Francisco. Remnant sea otter populations may have survived in the bay until 1840, when the Rancho Punta de Quentin was granted to Captain John B. R. Cooper, a sea captain from Boston, by Mexican Governor Juan Bautista Alvarado along with a license to hunt sea otters, reportedly then prevalent at the mouth of Corte Madera Creek.

Although the southern sea otter's range has continuously expanded from the remnant population of about 50 individuals in Big Sur since protection in 1911, however from 2007 to 2010, the otter population and its range contracted and since 2010 has made little progress. As of spring 2010, the northern boundary had moved from about Tunitas Creek to a point 2 km southeast of Pigeon Point, and the southern boundary has moved from approximately Coal Oil Point to Gaviota State Park. Recently, a toxin called microcystin, produced by a type of cyanobacteria (*Microcystis*), seems to be concentrated in the shellfish the otters eat, poisoning them. Cyanobacteria are found in stagnant freshwater enriched with nitrogen and phosphorus from septic tank and agricultural fertilizer runoff, and may be flushed into the ocean when streamflows are high in the rainy season. A record number of sea otter carcasses were found on California's coastline in 2010, with increased shark attacks an increasing component of the mortality. Great white sharks (*Carcharodon carcharias*) do not consume relatively fat-poor sea otters but shark-bitten carcasses have increased from 8% in the 1980s to 15% in the 1990s and to 30% in 2010 and 2011.

Otters were observed twice in Southern California in 2011, once near Laguna Beach and once at Zuniga Point Jetty, near San Diego. These are the first documented sightings of otters this far south in 30 years.

Oregon

The last native sea otter in Oregon was probably shot and killed in 1906. In 1970 and 1971, a total of 95 sea otters were transplanted from Amchitka Island, Alaska to the Southern Oregon coast. However, this translocation effort failed and otters soon again disappeared from the state.

In 2004, a lone male sea otter took up residence at Simpson Reef off of Cape Arago for six months. This male is thought to have originated from a colony in Washington, but disappeared after a coastal storm.

The most recent sighting of a sea otter off the Oregon coast took place on 18 February 2009, in Depoe Bay, Oregon. The lone male sea otter could have traveled from either California or Washington.

Ecology

Diet

Sea otters consume over 100 different prey species. In most of its range, the sea otter's diet consists almost exclusively of marine benthic invertebrates, including sea urchins, fat innkeeper worms, a variety of bivalves such as clams and mussels, abalone, other mollusks, crustaceans, and snails. Its prey ranges in size from tiny limpets and crabs to giant octopuses. Where prey such as sea urchins, clams, and abalone are present in a range of sizes, sea otters tend to select larger items over smaller ones of similar type. In California, they have been noted to ignore Pismo clams smaller than 3 inches (7 cm) across.

Sea otters keep kelp forests healthy by eating animals that graze on kelp

In a few northern areas, fish are also eaten. In studies performed at Amchitka Island in the 1960s, where the sea otter population was at carrying capacity, 50% of food found in sea otter stomachs was fish. The fish species were usually bottom-dwelling and sedentary or sluggish forms, such as *Hemilepidotus hemilepidotus* and family Tetraodontidae. However, south of Alaska on the North American coast, fish are a negligible or extremely minor part of the sea otter's diet. Contrary to popular depictions, sea otters rarely eat starfish, and any kelp that is consumed apparently passes through the sea otter's system undigested.

The individuals within a particular area often differ in their foraging methods and prey types, and tend to follow the same patterns as their mothers. The diet of local populations also changes over time, as sea otters can significantly deplete populations of highly preferred prey such as large sea urchins, and prey availability is also affected by other factors such as fishing by humans. Sea otters can thoroughly remove abalone from an area except for specimens in deep rock crevices, however, they never completely wipe out a prey species from an area. A 2007 Californian study demonstrated, in areas where food was relatively scarce, a wider variety of prey was consumed. Surprisingly, though, the diets of individuals were more specialized in these areas than in areas where food was plentiful.

As a Keystone Species

Sea otters are a classic example of a keystone species; their presence affects the ecosystem more profoundly than their size and numbers would suggest. They keep the population of certain benthic (sea floor) herbivores, particularly sea urchins, in check. Sea urchins graze on the lower stems of kelp, causing the kelp to drift away and die. Loss of the habitat and nutrients provided by kelp forests leads to profound cascade effects on the marine ecosystem. North Pacific areas that do not have sea otters often turn into urchin barrens, with abundant sea urchins and no kelp forest.

Remote areas of coastline, such as this area in California, sheltered the few remaining colonies of sea otters that survived the fur trade

Reintroduction of sea otters to British Columbia has led to a dramatic improvement in the health of coastal ecosystems, and similar changes have been observed as sea otter populations recovered in the Aleutian and Commander Islands and the Big Sur coast of California However, some kelp forest ecosystems in California have also thrived without sea otters, with sea urchin populations apparently controlled by other factors. The role of sea otters in maintaining kelp forests has been observed to be more important in areas of open coast than in more protected bays and estuaries.

In addition to promoting growth of kelp forests, sea otters can also have a profound effect in rocky areas that tend to be dominated by mussel beds. They remove mussels from rocks, liberating space for competitive species and thereby increasing the diversity of species in the area.

Predators

Predation of sea otters does occur, although it is not common. Many predators find the otter, with their pungent scent glands, distasteful. Young predators may kill an otter and not eat it. Leading mammalian predators of this species include orcas and sea lions; bald eagles also prey on pups by snatching them from the water surface. On land, young sea otters may face attack from bears and coyotes. In California, bites from sharks, particularly great white sharks, have been estimated to cause 10% of sea otter deaths and are one of the reasons the population has not expanded further north. The great white shark is believed to be their primary predator, and dead sea otters have been found with injuries from shark bites, although there is no evidence that sharks actually eat them. An exhibit at the San Diego Natural History Museum states that cat feces from urban runoff carry Toxoplasma gondii parasites to the ocean and kill sea otters.

Relationship with Humans

Fur Trade

Sea otters have the thickest fur of any mammal. Their beautiful fur is a main target for many hunters. Archaeological evidence indicates that for thousands of years, indigenous peoples have hunted sea otters for food and fur. Large-scale hunting, part of the Maritime Fur Trade, which would eventually kill approximately one million sea otters, began in the 18th century when hunters and traders began to arrive from all over the world to meet foreign demand for otter pelts, which were one of the world's most valuable types of fur.

Aleut men in Unalaska in 1896 used waterproof kayak gear and garments to hunt sea otters

In the early 18th century, Russians began to hunt sea otters in the Kuril Islands and sold them to the Chinese at Kyakhta. Russia was also exploring the far northern Pacific at this time, and sent Vitus Bering to map the Arctic coast and find routes from Siberia to North America. In 1741, on his second North Pacific voyage, Bering was shipwrecked off Bering Island in the Commander Islands, where he and many of his crew died. The surviving crew members, which included naturalist Georg Steller, discovered sea otters on the beaches of the island and spent the winter hunting sea otters and gambling with otter pelts. They returned to Siberia, having killed nearly 1,000 sea otters, and were able to command high prices for the pelts. Thus began what is sometimes called the "Great Hunt", which would continue for another hundred years. The Russians found the sea otter far more valuable than the sable skins that had driven and paid for most of their expansion across Siberia. If the sea otter pelts brought back by Bering's survivors had been sold at Kyakhta prices they would have paid for one tenth the cost of Bering's expedition. In 1775 at Okhotsk, sea otter pelts were worth 50–80 rubles as opposed to 2.5 rubles for sable.

Pelt sales (in thousands) in the London fur market – the drop beginning in the 1880s reflects dwindling sea otter populations.

Russian fur-hunting expeditions soon depleted the sea otter populations in the Commander Islands, and by 1745, they began to move on to the Aleutian Islands. The Russians initially traded with the Aleuts inhabitants of these islands for otter pelts, but later enslaved the Aleuts, taking women and children hostage and torturing and killing Aleut men to force them to hunt. Many Aleuts were either murdered by the Russians or died from diseases the hunters had introduced. The Aleut population was reduced, by the Russians' own estimate, from 20,000 to 2,000. By the 1760s, the Russians had reached Alaska. In 1799, Emperor Paul I consolidated the rival fur-hunting companies into the Russian-American Company, granting it an imperial charter and protection, and a monopoly over trade rights and territorial acquisition. Under Aleksandr I, the administration of the merchant-controlled company was transferred to the Imperial Navy, largely due to the alarming reports by naval officers of native abuse; in 1818, the indigenous peoples of Alaska were granted civil rights equivalent to a townsman status in the Russian Empire.

Other nations joined in the hunt in the south. Along the coasts of what is now Mexico and California, Spanish explorers bought sea otter pelts from Native Americans and sold them in Asia. In 1778, British explorer Captain James Cook reached Vancouver Island and bought sea otter furs from the First Nations people. When Cook's ship later stopped at a Chinese port, the pelts rapidly sold at high prices, and were soon known as "soft gold". As word spread, people from all over Europe and North America began to arrive in the Pacific Northwest to trade for sea otter furs.

Russian hunting expanded to the south, initiated by American ship captains, who subcontracted Russian supervisors and Aleut hunters in what are now Washington, Oregon, and California. Between 1803 and 1846, 72 American ships were involved in the otter hunt in California, harvesting an estimated 40,000 skins and tails, compared to only 13 ships of the Russian-American Company, which reported 5,696 otter skins taken between 1806 and 1846. In 1812, the Russians founded an agricultural settlement at what is now Fort Ross in northern California, as their southern headquarters. Eventually, sea otter populations became so depleted, commercial hunting was no longer viable. It had stopped the Aleutian Islands, by 1808, as a conservation measure imposed by the Russian-American Company. Further restrictions were ordered by the Company in 1834. When Russia sold Alaska to the United States in 1867, the Alaska population had recovered to over 100,000, but Americans resumed hunting and quickly extirpated the sea otter again. Prices rose as the species became rare. During the 1880s, a pelt brought $105 to $165 in the London market, but by 1903, a pelt could be worth as much as $1,125. In 1911, Russia, Japan, Great Britain (for Canada) and the United States signed the Treaty for the Preservation and Protection of Fur Seals, imposing a moratorium on the harvesting of sea otters. So few remained, perhaps only 1,000–2,000 individuals in the wild, that many believed the species would become extinct.

Recovery and Conservation

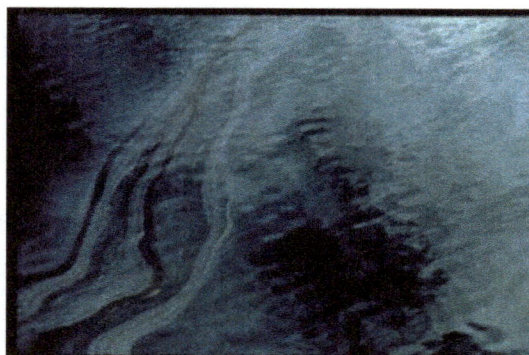

In the wake of the March 1989 Exxon Valdez oil spill, heavy sheens of oil covered large areas of Prince William Sound

During the 20th century, sea otter numbers rebounded in about two-thirds of their historic range, a recovery that is considered one of the greatest successes in marine conservation. However, the IUCN still lists the sea otter as an endangered species, and describes the significant threats to sea otters as oil pollution, predation by orcas, poaching, and conflicts with fisheries – sea otters can drown if entangled in fishing gear. The hunting of sea otters is no longer legal except for limited harvests by indigenous peoples in the United States. Poaching was a serious concern in the Russian Far East immediately after the collapse of the Soviet Union in 1991; however, it has declined significantly with stricter law enforcement and better economic conditions.

The most significant threat to sea otters is oil spills. They are particularly vulnerable, as they rely on their fur to keep warm. When their fur is soaked with oil, it loses its ability to retain air, and the animals can quickly die from hypothermia. The liver, kidneys, and lungs of sea otters also become damaged after they inhale oil or ingest it when grooming. The Exxon Valdez oil spill of 24 March 1989 killed thousands of sea otters in Prince William Sound, and as of 2006, the lingering oil in the area continues to affect the population. Describing the public sympathy for sea otters that developed from media coverage of the event, a U.S. Fish and Wildlife Service spokesperson wrote:

As a playful, photogenic, innocent bystander, the sea otter epitomized the role of victim ... cute and frolicsome sea otters suddenly in distress, oiled, frightened, and dying, in a losing battle with the oil.

The small geographic ranges of the sea otter populations in California, Washington, and British Columbia mean a single major spill could be catastrophic for that state or province. Prevention of oil spills and preparation for the rescue of otters in the event of one are major areas of focus for conservation efforts. Increasing the size and range of sea otter populations would also reduce the risk of an oil spill wiping out a population. However, because of the species' reputation for depleting shellfish resources, advocates for commercial, recreational, and subsistence shellfish harvesting have often opposed allowing the sea otter's range to increase, and there have even been instances of fishermen and others illegally killing them.

Sea otters in the Olympic Coast National Marine Sanctuary – note the unusual shape of the hind feet, in which the outer toes are longest

In the Aleutian Islands, a massive and unexpected disappearance of sea otters has occurred in recent decades. In the 1980s, the area was home to an estimated 55,000 to 100,000 sea otters, but the population fell to around 6,000 animals by 2000. The most widely accepted, but still controversial, hypothesis is that killer whales have been eating the otters. The pattern of disappearances is consistent with a rise in predation, but there has been no direct evidence of orcas preying on sea otters to any significant extent.

Another area of concern is California, where recovery began to fluctuate or decline in the late 1990s. Unusually high mortality rates amongst adult and subadult otters, particularly females, have been reported. Necropsies of dead sea otters indicate diseases, particularly *Toxoplasma gondii* and acanthocephalan parasite infections, are major causes of sea otter mortality in California. The *Toxoplasma gondii* parasite, which is often fatal to sea otters, is carried by wild and domestic cats and may be transmitted by domestic cat droppings flushed into the ocean via sewage systems. Although disease has clearly contributed to the deaths of many of California's sea otters, it is not known why the California population is apparently more affected by disease than populations in other areas.

Sea otter habitat is preserved through several protected areas in the United States, Russia and Canada. In marine protected areas, polluting activities such as dumping of waste and oil drilling are typically prohibited. An estimated 1,200 sea otters live within the Monterey Bay National Marine Sanctuary, and more than 500 live within the Olympic Coast National Marine Sanctuary.

Economic Impact

Some of the sea otter's preferred prey species, particularly abalone, clams, and crabs, are also food sources for humans. In some areas, massive declines in shellfish harvests have been blamed on the sea otter, and intense public debate has taken place over how to manage the competition between sea otters and humans for seafood.

Sea otters like this one near Moss Landing are a tourist attraction in the Monterey Bay area in California

The debate is complicated because sea otters have sometimes been held responsible for declines of shellfish stocks that were more likely caused by overfishing, disease, pollution, and seismic activity. Shellfish declines have also occurred in many parts of the North American Pacific coast that do not have sea otters, and conservationists sometimes note the existence of large concentrations of shellfish on the coast is a recent development resulting from the fur trade's near-extirpation of the sea otter. Although many factors affect shellfish stocks, sea otter predation can deplete a fishery to the point where it is no longer commercially viable. Scientists agree that sea otters and abalone fisheries cannot exist in the same area, and the same is likely true for certain other types of shellfish, as well.

Many facets of the interaction between sea otters and the human economy are not as immediately felt. Sea otters have been credited with contributing to the kelp harvesting industry via their well-known role in controlling sea urchin populations; kelp is used in the production of diverse food and pharmaceutical products. Although human divers harvest red sea urchins both for food and to protect the kelp, sea otters hunt more sea urchin species and are more consistently effective in controlling these populations. The health of the kelp forest ecosystem is significant in nurturing populations of fish, including commercially important fish species. In some areas, sea otters are popular tourist attractions, bringing visitors to local hotels, restaurants, and sea otter-watching expeditions.

Roles in Human Cultures

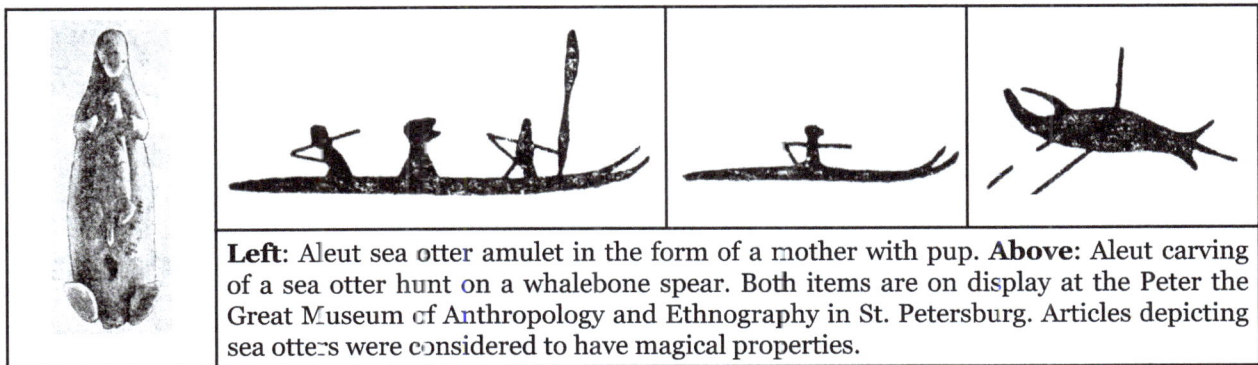

Left: Aleut sea otter amulet in the form of a mother with pup. **Above**: Aleut carving of a sea otter hunt on a whalebone spear. Both items are on display at the Peter the Great Museum of Anthropology and Ethnography in St. Petersburg. Articles depicting sea otters were considered to have magical properties.

For many maritime indigenous cultures throughout the North Pacific, especially the Ainu in the Kuril Islands, the Koryaks and Itelmen of Kamchatka, the Aleut in the Aleutian Islands, the Haida of Haida Gwaii and a host of tribes on the Pacific coast of North America, the sea otter has played an important role as a cultural, as well as material, resource. In these cultures, many of which have strongly animist traditions full of legends and stories in which many aspects of the natural world are associated with spirits, the sea otter was considered particularly kin to humans. The Nuu-chah-nulth, Haida, and other First Nations of coastal British Columbia used the warm and luxurious pelts as chiefs' regalia. Sea otter pelts were given in potlatches to mark coming-of-age ceremonies, weddings, and funerals. The Aleuts carved sea otter bones for use as ornaments and in games, and used powdered sea otter baculum as a medicine for fever.

Sea otters at the Lisbon Oceanarium show their flexibility when grooming

Among the Ainu, the otter is portrayed as an occasional messenger between humans and the creator. The sea otter is a recurring figure in Ainu folklore. A major Ainu epic, the *Kutune Shirka*, tells the tale of wars and struggles over a golden sea otter. Versions of a widespread Aleut legend tell of lovers or despairing women who plunge into the sea and become otters. These links have been associated with the many human-like behavioral features of the sea otter, including apparent playfulness, strong mother-pup bonds and tool use, yielding to ready anthropomorphism. The beginning of commercial exploitation had a great impact on the human, as well as animal, populations the Ainu and Aleuts have been displaced or their numbers are dwindling, while the coastal tribes of North America, where the otter is in any case greatly depleted, no longer rely as intimately on sea mammals for survival.

Since the mid-1970s, the beauty and charisma of the species have gained wide appreciation, and the sea otter has become an icon of environmental conservation. The round, expressive face and soft, furry body of the sea otter are depicted in a wide variety of souvenirs, postcards, clothing, and stuffed toys.

H1N1 Host

According to the U.S. Geological Survey and the CDC, northern sea otters, off the coast of Washington state, are infected with the H1N1 flu virus and "may be a newly identified animal host of influenza viruses".

Aquariums and Zoos

Sea otters can do well in captivity, and are featured in over 40 public aquariums and zoos. The Seattle Aquarium became the first institution to raise sea otters from conception to adulthood with the birth of Tichuk in 1979, followed by three more pups in the early 1980s. In 2007, a YouTube video of two sea otters holding paws drew 1.5 million viewers in two weeks, and had over 20 million views as of January 2015. Filmed five years previously at the Vancouver Aquarium, it was YouTube's most popular animal video at the time, although it has since been surpassed. The lighter-colored otter in the video is Nyac, a survivor of the 1989 *Exxon Valdez* oil spill. Nyac died in September 2008, at the age of 20. Milo, the darker one, died of lymphoma in January, 2012

Polar Bear

The polar bear (*Ursus maritimus*) is a carnivorous bear whose native range lies largely within the Arctic Circle, encompassing the Arctic Ocean, its surrounding seas and surrounding land masses. It is a large bear, approximately the same size as the omnivorous Kodiak bear (*Ursus arctos middendorffi*). A boar (adult male) weighs around 350–700 kg (772–1,543 lb), while a sow (adult female) is about half that size. Although it is the sister species of the brown bear, it has evolved to occupy a narrower ecological niche, with many body characteristics adapted for cold temperatures, for moving across snow, ice and open water, and for hunting seals, which make up most of its diet. Although most polar bears are born on land, they spend most of their time on the sea ice. Their scientific name means "maritime bear" and derives from this fact. Polar bears hunt their preferred food of seals from the edge of sea ice, often living off fat reserves when no sea ice is present. Because of their dependence on the sea ice, polar bears are classified as marine mammals.

Because of expected habitat loss caused by climate change, the polar bear is classified as a vulnerable species, and at least three of the nineteen polar bear subpopulations are currently in decline. For decades, large-scale hunting raised international concern for the future of the species, but populations rebounded after controls and quotas began to take effect. For thousands of years, the polar bear has been a key figure in the material, spiritual, and cultural life of circumpolar peoples, and polar bears remain important in their cultures.

Naming and Etymology

Constantine John Phipps was the first to describe the polar bear as a distinct species in 1774. He chose the scientific name *Ursus maritimus*, the Latin for 'maritime bear', due to the animal's native habitat. The Inuit refer to the animal as *nanook*.

The polar bear was previously considered to be in its own genus, *Thalarctos*. However, evidence of hybrids between polar bears and brown bears, and of the recent evolutionary divergence of the two species, does not support the establishment of this separate genus, and the accepted scientific name is now therefore *Ursus maritimus*, as Phipps originally proposed.

Taxonomy and Evolution

Polar bears have evolved adaptations for Arctic life, for example, large furry feet and short, sharp, stocky claws give them good traction on ice.

The bear family, Ursidae, is thought to have split off from other carnivorans about 38 million years ago. The Ursinae subfamily originated approximately 4.2 million years ago. The oldest known polar bear fossil is a 130,000 to 110,000-year-old jaw bone, found on Prince Charles Foreland in 2004. Fossils show that between 10,000 and 20,000 years ago, the polar bear's molar teeth changed significantly from those of the brown bear. Polar bears are thought to have diverged from a population of brown bears that became isolated during a period of glaciation in the Pleistocene or from the eastern part of Siberia, (from Kamchatka and the Kolym Peninsula).

The evidence from DNA analysis is more complex. The mitochondrial DNA (mtDNA) of the polar bear diverged from the brown bear, *Ursus arctos*, roughly 150,000 years ago. Further, some clades of brown bear, as assessed by their mtDNA, are more closely related to polar bears than to other brown bears, meaning that the polar bear might not be considered a species under some species concepts. The mtDNA of extinct Irish brown bears is particularly close to polar bears. A comparison of the nuclear genome of polar bears with that of brown bears revealed a different pattern, the two forming genetically distinct clades that diverged approximately 603,000 years ago, although the latest research is based on analysis of the complete genomes (rather than just the mitochondria or partial nuclear genomes) of polar and brown bears, and establishes the divergence of polar and brown bears at 400,000 years ago.

However, the two species have mated intermittently for all that time, most likely coming into contact with each other during warming periods, when polar bears were driven onto land and brown bears migrated northward. Most brown bears have about 2 percent genetic material from polar bears, but one population, the ABC Islands bears has between 5 percent and 10 percent polar bear genes, indicating more frequent and recent mating. Polar bears can breed with brown bears to produce fertile grizzly–polar bear hybrids, rather than indicating that they have only recently diverged, the new evidence suggests more frequent mating has continued over a longer period of time, and thus the two bears remain genetically similar. However, because neither species can survive long in the other's ecological niche, and because they have different morphology, metabolism, social and feeding behaviours, and other phenotypic characteristics, the two bears are generally classified as separate species.

When the polar bear was originally documented, two subspecies were identified: *Ursus maritimus maritimus* by Constantine J. Phipps in 1774, and *Ursus maritimus marinus* by Peter Simon Pallas in 1776. This distinction has since been invalidated. One alleged fossil subspecies has been identified: *Ursus maritimus tyrannus* became extinct during the Pleistocene. *U.m. tyrannus* was significantly larger than the living subspecies. However, recent reanalysis of the fossil suggests that it was actually a type of brown bear.

Population and Distribution

Polar bears investigate the submarine *USS Honolulu* 450 kilometres (280 mi) from the North Pole

The polar bear is found in the Arctic Circle and adjacent land masses as far south as Newfoundland. Due to the absence of human development in its remote habitat, it retains more of its original range than any other extant carnivore. While they are rare north of 88°, there is evidence that they range all the way across the Arctic, and as far south as James Bay in Canada. Their southernmost range is near the boundary between the subarctic and humid continental climate zones. They can occasionally drift widely with the sea ice, and there have been anecdotal sightings as far south as Berlevåg on the Norwegian mainland and the Kuril Islands in the Sea of Okhotsk. It is difficult to estimate a global population of polar bears as much of the range has been poorly studied; however, biologists use a working estimate of about 20–25,000 or 22–31,000 polar bears worldwide.

There are 19 generally recognized, discrete subpopulations, though polar bears are thought to exist only in low densities in the area of the Arctic Basin. The subpopulations display seasonal fidelity to particular areas, but DNA studies show that they are not reproductively isolated. The thirteen North American subpopulations range from the Beaufort Sea south to Hudson Bay and east to Baffin Bay in western Greenland and account for about 54% of the global population. The Eurasian population is broken up into the eastern Greenland, Barents Sea, Kara Sea, Laptev Sea, and Chukchi Sea subpopulations, though there is considerable uncertainty about the structure of these populations due to limited mark and recapture data.

The range includes the territory of five nations: Denmark (Greenland), Norway (Svalbard), Russia, the United States (Alaska) and Canada. These five nations are the signatories of the International Agreement on the Conservation of Polar Bears, which mandates cooperation on research and conservation efforts throughout the polar bear's range.

Modern methods of tracking polar bear populations have been implemented only since the mid-1980s, and are expensive to perform consistently over a large area. The most accurate counts require flying a helicopter in the Arctic climate to find polar bears, shooting a tranquilizer dart at the bear to sedate it, and then tagging the bear. In Nunavut, some Inuit have reported increases in bear sightings around human settlements in recent years, leading to a belief that populations are increasing. Scientists have responded by noting that hungry bears may be congregating around human settlements, leading to the illusion that populations are higher than they actually are. The Polar Bear Specialist Group of the IUCN Species Survival Commission takes the position that "estimates of subpopulation size or sustainable harvest levels should not be made solely on the basis of traditional ecological knowledge without supporting scientific studies."

Of the 19 recognized polar bear subpopulations, three are declining, six are stable, one is increasing, and nine have insufficient data, as of 2014.

Habitat

The polar bear is a marine mammal because it spends many months of the year at sea. However, it is the only living marine mammal with powerful, large limbs and feet that allow them to cover miles on foot and run on land. Its preferred habitat is the annual sea ice covering the waters over the continental shelf and the Arctic inter-island archipelagos. These areas, known as the "Arctic ring of life", have high biological productivity in comparison to the deep waters of the high Arctic. The polar bear tends to frequent areas where sea ice meets water, such as polynyas and leads (temporary stretches of open water in Arctic ice), to hunt the seals that make up most of its diet.

Freshwater is limited in these environments because it is either locked up in snow or saline. Polar bears are able to produce water through the metabolism of fats found in seal blubber. Polar bears are therefore found primarily along the perimeter of the polar ice pack, rather than in the Polar Basin close to the North Pole where the density of seals is low.

Polar bear jumping on fast ice

Annual ice contains areas of water that appear and disappear throughout the year as the weather changes. Seals migrate in response to these changes, and polar bears must follow their prey. In Hudson Bay, James Bay, and some other areas, the ice melts completely each summer (an event often referred to as "ice-floe breakup"), forcing polar bears to go onto land and wait through the months until the next freeze-up. In the Chukchi and Beaufort seas, polar bears retreat each summer to the ice further north that remains frozen year-round.

Physical Characteristics

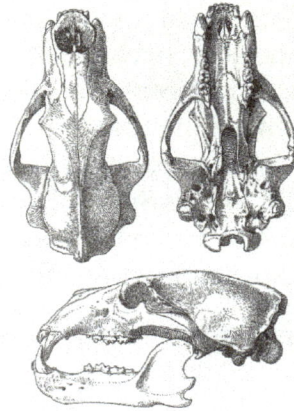

Skull, as illustrated by N. N. Kondakov

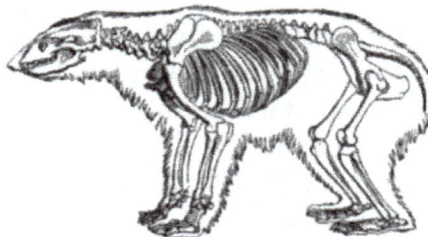

Polar bear skeleton

The only other bear of a similar size to the polar bear is the Kodiak bear, which is a subspecies of brown bear. Adult male polar bears weigh 350–700 kg (772–1,543 lb) and measure 2.4–3 metres (7 ft 10 in–9 ft 10 in) in total length. The Guinness Book of World Records listed the average male as having a body mass of 385 to 410 kg (849 to 904 lb) and a shoulder height of 133 cm (4 ft 4 in), slightly smaller than the average male Kodiak bears. Around the Beaufort Sea, however, mature males reportedly average 450 kg (992 lb). Adult females are roughly half the size of males and normally weigh 150–250 kg (331–551 lb), measuring 1.8–2.4 metres (5 ft 11 in–7 ft 10 in) in length. Elsewhere, a slightly larger estimated average weight of 260 kg (573 lb) was claimed for adult females. When pregnant, however, females can weigh as much as 500 kg (1,102 lb). The polar bear is among the most sexually dimorphic of mammals, surpassed only by the pinnipeds such as elephant seals. The largest polar bear on record, reportedly weighing 1,002 kg (2,209 lb), was a male shot at Kotzebue Sound in northwestern Alaska in 1960. This specimen, when mounted, stood 3.39 m (11 ft 1 in) tall on its hindlegs. The shoulder height of an adult polar bear is 122 to 160 cm (4 ft 0 in to 5 ft 3 in). While all bears are short-tailed, the polar bear's tail is relatively the shortest amongst living bears, ranging from 7 to 13 cm (2.8 to 5.1 in) in length.

Compared with its closest relative, the brown bear, the polar bear has a more elongated body build and a longer skull and nose. As predicted by Allen's rule for a northerly animal, the legs are stocky and the ears and tail are small. However, the feet are very large to distribute load when walking on snow or thin ice and to provide propulsion when swimming; they may measure 30 cm (12 in) across in an adult. The pads of the paws are covered with small, soft papillae (dermal bumps), which provide traction on the ice. The polar bear's claws are short and stocky compared to those of the brown bear, perhaps to serve the former's need to grip heavy prey and ice. The claws are deeply scooped on the underside to assist in digging in the ice of the natural habitat. Research of injury patterns in polar bear forelimbs found injuries to the right forelimb to be more frequent than those to the left, suggesting, perhaps, right-handedness. Unlike the brown bear, polar bears in captivity are rarely overweight or particularly large, possibly as a reaction to the warm conditions of most zoos.

The 42 teeth of a polar bear reflect its highly carnivorous diet. The cheek teeth are smaller and more jagged than in the brown bear, and the canines are larger and sharper. The dental formula is 3.1.4.23.1.4.3

Polar bears are superbly insulated by up to 10 cm (4 in) of adipose tissue, their hide and their fur; they overheat at temperatures above 10 °C (50 °F), and are nearly invisible under infrared photography. Polar bear fur consists of a layer of dense underfur and an outer layer of guard hairs, which appear white to tan but are actually transparent. The guard hair is 5–15 cm (2–6 in) over most of the body. Polar bears gradually moult from May to August, but, unlike other Arctic mammals, they do not shed their coat for a darker shade to camouflage themselves in the summer conditions. The hollow guard hairs of a polar bear coat were once thought to act as fiber-optic tubes to conduct light to its black skin, where it could be absorbed; however, this hypothesis was disproved by a study in 1998.

The white coat usually yellows with age. When kept in captivity in warm, humid conditions, the fur may turn a pale shade of green due to algae growing inside the guard hairs. Males have significantly longer hairs on their forelegs, which increase in length until the bear reaches 14 years of age. The male's ornamental foreleg hair is thought to attract females, serving a similar function to the lion's mane.

The polar bear has an extremely well developed sense of smell, being able to detect seals nearly 1.6 km (1 mi) away and buried under 1 m (3 ft) of snow. Its hearing is about as acute as that of a human, and its vision is also good at long distances.

The polar bear is an excellent swimmer and often will swim for days. One bear swam continuously for 9 days in the frigid Bering Sea for 400 mi (687 km) to reach ice far from land. She then travelled another 1,100 mi (1,800 km). During the swim, the bear lost 22% of her body mass and her yearling cub died. With its body fat providing buoyancy, the bear swims in a dog paddle fashion using its large forepaws for propulsion. Polar bears can swim 10 km/h (6 mph). When walking, the polar bear tends to have a lumbering gait and maintains an average speed of around 5.6 km/h (3.5 mph). When sprinting, they can reach up to 40 km/h (25 mph).

Life History and Behaviour

Subadult polar bear males frequently play-fight. During the mating season, actual fighting is intense and often leaves scars or broken teeth.

Unlike grizzly bears, polar bears are not territorial. Although stereotyped as being voraciously aggressive, they are normally cautious in confrontations, and often choose to escape rather than fight. Satiated polar bears rarely attack humans unless severely provoked. However, due to their lack of prior human interaction, hungry polar bears are extremely unpredictable, fearless towards people and are known to kill and sometimes eat humans. Many attacks by brown bears are the result of surprising the animal, which is not the case with the polar bear. Polar bears are stealth hunters, and the victim is often unaware of the bear's presence until the attack is underway. Whereas brown bears often maul a person and then leave, polar bear attacks are more likely to be predatory and are almost always fatal. However, due to the very small human population around the Arctic, such attacks are rare. Michio Hoshino, a Japanese wildlife photographer, was once pursued briefly by a hungry male polar bear in northern Alaska. According to Hoshino, the bear started running but Hoshino made it to his truck. The bear was able to reach the truck and tore one of the doors off the truck before Hoshino was able to drive off.

In general, adult polar bears live solitary lives. Yet, they have often been seen playing together for hours at a time and even sleeping in an embrace, and polar bear zoologist Nikita Ovsianikov has described adult males as having "well-developed friendships." Cubs are especially playful as well.

Among young males in particular, play-fighting may be a means of practicing for serious competition during mating seasons later in life. Polar bears have a wide range of vocalisations, including bellows, roars, growls, chuffs and purrs.

In 1992, a photographer near Churchill took a now widely circulated set of photographs of a polar bear playing with a Canadian Eskimo Dog (*Canis lupus familiaris*) a tenth of its size. The pair wrestled harmlessly together each afternoon for ten days in a row for no apparent reason, although the bear may have been trying to demonstrate its friendliness in the hope of sharing the kennel's food. This kind of social interaction is uncommon; it is far more typical for polar bears to behave aggressively towards dogs.

Hunting and Diet

Long muzzle and neck of the polar bear help it to search in deep holes for seals, while powerful hindquarters enable it to drag massive prey

The polar bear is the most carnivorous member of the bear family, and throughout most of its range, its diet primarily consists of ringed (*Pusa hispida*) and bearded seals (*Erignathus barbatus*). The Arctic is home to millions of seals, which become prey when they surface in holes in the ice in order to breathe, or when they haul out on the ice to rest. Polar bears hunt primarily at the interface between ice, water, and air; they only rarely catch seals on land or in open water.

The polar bear's most common hunting method is called *still-hunting*: The bear uses its excellent sense of smell to locate a seal breathing hole, and crouches nearby in silence for a seal to appear. The bear may lay in wait for several hours. When the seal exhales, the bear smells its breath, reaches into the hole with a forepaw, and drags it out onto the ice. The polar bear kills the seal by biting its head to crush its skull. The polar bear also hunts by stalking seals resting on the ice: Upon spotting a seal, it walks to within 90 m (100 yd), and then crouches. If the seal does not notice, the bear creeps to within 9 to 12 m (30 to 40 ft) of the seal and then suddenly rushes forth to attack. A third hunting method is to raid the birth lairs that female seals create in the snow.

A widespread legend tells that polar bears cover their black noses with their paws when hunting. This behaviour, if it happens, is rare – although the story exists in the oral history of northern peoples and in accounts by early Arctic explorers, there is no record of an eyewitness account of the behaviour in recent decades.

Polar bear feeding on a bearded seal

Mature bears tend to eat only the calorie-rich skin and blubber of the seal, which are highly digestible, whereas younger bears consume the protein-rich red meat. Studies have also photographed polar bears scaling near-vertical cliffs, to eat birds' chicks and eggs. For subadult bears, which are independent of their mother but have not yet gained enough experience and body size to successfully hunt seals, scavenging the carcasses from other bears' kills is an important source of nutrition. Subadults may also be forced to accept a half-eaten carcass if they kill a seal but cannot defend it from larger polar bears. After feeding, polar bears wash themselves with water or snow.

Although polar bears are extraordinarily powerful, its primary prey species, the ringed seal, is much smaller than itself, and many of the seals hunted are pups rather than adults. Ringed seals are born weighing 5.4 kg (12 lb) and grown to an estimated average weight of only 60 kg (130 lb). They also in places prey heavily upon the harp seal (*Pusa groenlandica*) or the harbor seal. The bearded seal, on the other hand, can be nearly the same size as the bear itself, averaging 270 kg (600 lb). Adult male bearded seals, at 350 to 500 kg (770 to 1,100 lb) are too large for a female bear to overtake, and so are potential prey only for mature male bears. Large males also occasionally attempt to hunt and kill even larger prey items. It can kill an adult walrus (*Odobenus rosmarus*), although this is rarely attempted. At up to 2,000 kg (4,400 lb) and a typical adult mass range of 600 to 1,500 kg (1,300 to 3,300 lb), a walrus can be more than twice the bear's weight, and has up to 1-metre (3 ft)-long ivory tusks that can be used as formidable weapons. A polar bear may charge a group of walruses, with the goal of separating a young, infirm, or injured walrus from the pod. They will even attack adult walruses when their diving holes have frozen over or intercept them before they can get back to the diving hole in the ice. Yet, polar bears will very seldom attack full-grown adult walruses, with the largest male walrus probably invulnerable unless otherwise injured or incapacitated. Since an attack on a walrus tends to be an extremely protracted and exhausting venture, bears have been known to back down from the attack after making the initial injury to the walrus. Polar bears have also been seen to prey on beluga whales (*Delphinapterus leucas*) and narwhals (*Monodon monoceros*), by swiping at them at breathing holes. The whales are of similar size to the walrus and nearly as difficult for the bear to subdue. Most terrestrial animals in the Arctic can outrun the polar bear on land as polar bears overheat quickly, and most marine animals the bear encounters can outswim it. In some areas, the polar bear's diet is supplemented by walrus calves and by the carcasses of dead adult walruses or whales, whose blubber is readily devoured even when rotten. Polar bears sometimes like to go fishing where they swim underwater to catch fish like the Arctic charr or the fourhorn sculpin.

Some characteristic postures: 1. at rest; 2. assessing a situation; 3.when feeding

With the exception of pregnant females, polar bears are active year-round, although they have a vestigial hibernation induction trigger in their blood. Unlike brown and black bears, polar bears are capable of fasting for up to several months during late summer and early fall, when they cannot hunt for seals because the sea is unfrozen. When sea ice is unavailable during summer and early autumn, some populations live off fat reserves for months at a time, as polar bears do not 'hibernate' any time of the year.

Being both curious animals and scavengers, polar bears investigate and consume garbage where they come into contact with humans. Polar bears may attempt to consume almost anything they can find, including hazardous substances such as styrofoam, plastic, car batteries, ethylene glycol, hydraulic fluid, and motor oil. The dump in Churchill, Manitoba was closed in 2006 to protect bears, and waste is now recycled or transported to Thompson, Manitoba.

Dietary Flexibility

Although seal predation is the primary and an indispensable way of life for most polar bears, when alternatives are present they are quite flexible. Polar bears will consume a wide variety of other wild foods, including muskox (*Ovibos moschatus*), reindeer (*Rangifer tarandus*), birds, eggs, rodents, crabs, other crustaceans and other polar bears. They may also eat plants, including berries, roots, and kelp; however, none of these are a significant part of their diet, except that beachcast marine mammal carcasses are an exception. When stalking land animals, such as muskox, reindeer, and even willow ptarmigan (*Lagopus lagopus*), polar bears appear to make use of vegetative cover and wind direction to bring them as close to their prey as possible before attacking. Polar bears have been observed to hunt the small Svalbard reindeer (*R. t. platyrhynchus*), which weigh only 40 to 60 kg (90 to 130 lb) as adults, as well as the barren-ground caribou (*R. t. groenlandicus*), which is about twice as heavy as that. Adult muskox, which can weigh 450 kg (1,000 lb) or more, are a more formidable quarry. Although ungulates are not typical prey, the killing of one during the summer months can greatly increase the odds of survival during that lean period. Like the brown bear, most ungulate prey of polar bears is likely to be young, sickly or injured specimens rather than healthy adults. The polar bear's biology is specialized to require large amounts of fat from marine mammals, and it cannot derive sufficient caloric intake from terrestrial food.

In their southern range, especially near Hudson Bay and James Bay, Canadian polar bears endure all summer without sea ice to hunt from. Here, their food ecology shows their dietary flexibility. They still manage to consume some seals, but they are food-deprived in summer as only marine mammal carcasses are an important alternative without sea ice, especially carcasses of the beluga whale. These alternatives may reduce the rate of weight loss of bears when on land. One scientist found that 71% of the Hudson Bay bears had fed on seaweed (marine algae) and that about half were feeding on birds like sea ducks, especially the oldsquaw (53%), common eider, long-tailed duck or dovekie by swimming underwater to catch them. They were also diving to feed on blue mussels and other underwater food sources like the green sea urchin. 24% had eaten moss recently, 19% had consumed grass, 34% had eaten black crowberry and about half had consumed willows. This study illustrates the polar bear's dietary flexibility but it does not represent its life history elsewhere. Most polar bears elsewhere will never have access to these alternatives, except for the marine mammal carcasses that are important wherever they occur.

In Svalbard, polar bears were observed to kill white-beaked dolphins during spring, when the dolphins were trapped in the sea ice. The bears then proceeded to cache the carcasses, which remained and were eaten during the ice-free summer and autumn.

Reproduction and Lifecycle

Cubs are born helpless and typically nurse for two and a half years

Courtship and mating take place on the sea ice in April and May, when polar bears congregate in the best seal hunting areas. A male may follow the tracks of a breeding female for 100 km (60 mi) or more, and after finding her engage in intense fighting with other males over mating rights, fights that often result in scars and broken teeth. Polar bears have a generally polygynous mating system; recent genetic testing of mothers and cubs, however, has uncovered cases of litters in which cubs have different fathers. Partners stay together and mate repeatedly for an entire week; the mating ritual induces ovulation in the female.

After mating, the fertilized egg remains in a suspended state until August or September. During these four months, the pregnant female eats prodigious amounts of food, gaining at least 200 kg (440 lb) and often more than doubling her body weight.

Maternity Denning and Early Life

Mother and cub on Svalbard

When the ice floes break up in the fall, ending the possibility of hunting, each pregnant female digs a *maternity den* consisting of a narrow entrance tunnel leading to one to three chambers. Most maternity dens are in snowdrifts, but may also be made underground in permafrost if it is not sufficiently cold yet for snow. In most subpopulations, maternity dens are situated on land a few kilometers from the coast, and the individuals in a subpopulation tend to reuse the same denning areas each year. The polar bears that do not den on land make their dens on the sea ice. In the den, she enters a dormant state similar to hibernation. This hibernation-like state does not consist of continuous sleeping; however, the bear's heart rate slows from 46 to 27 beats per minute. Her body temperature does not decrease during this period as it would for a typical mammal in hibernation.

Between November and February, cubs are born blind, covered with a light down fur, and weighing less than 0.9 kg (2.0 lb), but in captivity they might be delivered in the earlier months. The earliest recorded birth of polar bears in captivity was on 11 October 2011 in the Toronto Zoo. On average, each litter has two cubs. The family remains in the den until mid-February to mid-April, with the mother maintaining her fast while nursing her cubs on a fat-rich milk. By the time the mother breaks open the entrance to the den, her cubs weigh about 10 to 15 kilograms (22 to 33 lb). For about 12 to 15 days, the family spends time outside the den while remaining in its vicinity, the mother grazing on vegetation while the cubs become used to walking and playing. Then they begin the long walk from the denning area to the sea ice, where the mother can once again catch seals. Depending on the timing of ice-floe breakup in the fall, she may have fasted for up to eight months. During this time, cubs playfully imitate the mother's hunting methods in preparation for later life.

Female polar bears are noted for both their affection towards their offspring, and their valor in protecting them. Multiple cases of adoption of wild cubs have been confirmed by genetic testing. Adult male bears occasionally kill and eat polar bear cubs. As of 2006, in Alaska, 42% of cubs were reaching 12 months of age, down from 65% in 1991. In most areas, cubs are weaned at two and a half years of age, when the mother chases them away or abandons them. The Western Hudson Bay subpopulation is unusual in that its female polar bears sometimes wean their cubs at only one and a half years. This was the case for 40% of cubs there in the early 1980s; however by the 1990s, fewer than 20% of cubs were weaned this young. After the mother leaves, sibling cubs sometimes travel and share food together for weeks or months.

Later Life

Females begin to breed at the age of four years in most areas, and five years in the Beaufort Sea area. Males usually reach sexual maturity at six years; however, as competition for females is fierce, many do not breed until the age of eight or ten. A study in Hudson Bay indicated that both the reproductive success and the maternal weight of females peaked in their mid-teens.

Polar bears appear to be less affected by infectious diseases and parasites than most terrestrial mammals. Polar bears are especially susceptible to *Trichinella*, a parasitic roundworm they contract through cannibalism, although infections are usually not fatal. Only one case of a polar bear with rabies has been documented, even though polar bears frequently interact with Arctic foxes, which often carry rabies. Bacterial leptospirosis and *Morbillivirus* have been recorded. Polar bears sometimes have problems with various skin diseases that may be caused by mites or other parasites.

Life Expectancy

Polar bears rarely live beyond 25 years. The oldest wild bears on record died at age 32, whereas the oldest captive was a female who died in 1991, age 43. The causes of death in wild adult polar bears are poorly understood, as carcasses are rarely found in the species's frigid habitat. In the wild, old polar bears eventually become too weak to catch food, and gradually starve to death. Polar bears injured in fights or accidents may either die from their injuries or become unable to hunt effectively, leading to starvation.

Ecological Role

The polar bear is the apex predator within its range, and is a keystone species for the Arctic. Several animal species, particularly Arctic foxes (*Vulpes lagopus*) and glaucous gulls (*Larus hyperboreus*), routinely scavenge polar bear kills.

The relationship between ringed seals and polar bears is so close that the abundance of ringed seals in some areas appears to regulate the density of polar bears, while polar bear predation in turn regulates density and reproductive success of ringed seals. The evolutionary pressure of polar bear predation on seals probably accounts for some significant differences between Arctic and Antarctic seals. Compared to the Antarctic, where there is no major surface predator, Arctic seals use more breathing holes per individual, appear more restless when hauled out on the ice, and rarely defecate on the ice. The baby fur of most Arctic seal species is white, presumably to provide camouflage from predators, whereas Antarctic seals all have dark fur at birth.

Brown bears tend to dominate polar bears in disputes over carcasses, and dead polar bear cubs have been found in brown bear dens. Wolves are rarely encountered by polar bears, though there are two records of Arctic wolf (*Canis lupus arctos*) packs killing polar bear cubs. A rather unlikely killer of a grown polar bear has reportedly included a wolverine (*Gulo gulo*), anecdotely reported to have suffocated a bear in a zoo with a bite to the throat during a conflict. Polar bears are sometimes the host of arctic mites such as *Alaskozetes antarcticus*.

Long-distance Swimming and Diving

Researchers tracked 52 sows in the southern Beaufort Sea off Alaska with GPS system collars; no boars were involved in the study due to males' necks being too thick for the GPS-equipped collars. Fifty long-distance swims were recorded; the longest at 354 kilometres (220 mi), with an average of 155 kilometres (96 mi). The length of these swims ranged from most of a day to ten days. Ten of the sows had a cub swim with them and after a year, six cubs survived. The study did not determine if the others lost their cubs before, during, or some time after their long swims. Researchers do not know whether or not this is a new behaviour; before polar ice shrinkage, they opined that there was probably neither the need nor opportunity to swim such long distances.

The polar bear may swim underwater for up to three minutes to approach seals on shore or on ice floes.

Hunting

Indigenous People

Skins of hunted polar bears

Polar bears have long provided important raw materials for Arctic peoples, including the Inuit, Yupik, Chukchi, Nenets, Russian Pomors and others. Hunters commonly used teams of dogs to distract the bear, allowing the hunter to spear the bear or shoot it with arrows at closer range. Almost all parts of captured animals had a use. The fur was used in particular to make trousers and, by the Nenets, to make galoshes-like outer footwear called *tobok*; the meat is edible, despite some risk of trichinosis; the fat was used in food and as a fuel for lighting homes, alongside seal and whale blubber; sinews were used as thread for sewing clothes; the gallbladder and sometimes heart were dried and powdered for medicinal purposes; the large canine teeth were highly valued as talismans. Only the liver was not used, as its high concentration of vitamin A is poisonous. Hunters make sure to either toss the liver into the sea or bury it in order to spare their dogs from potential poisoning. Traditional subsistence hunting was on a small enough scale to not significantly affect polar bear populations, mostly because of the sparseness of the human population in polar bear habitat.

History of Commercial Harvest

In Russia, polar bear furs were already being commercially traded in the 14th century, though it was of low value compared to Arctic fox or even reindeer fur. The growth of the human population in the Eurasian Arctic in the 16th and 17th century, together with the advent of firearms and increasing trade, dramatically increased the harvest of polar bears. However, since polar bear fur has always played a marginal commercial role, data on the historical harvest is fragmentary. It is known, for example, that already in the winter of 1784/1785 Russian Pomors on Spitsbergen harvested 150 polar bears in Magdalenefjorden. In the early 20th century, Norwegian hunters were harvesting 300 bears per year at the same location. Estimates of total historical harvest suggest that from the beginning of the 18th century, roughly 400 to 500 animals were being harvested annually in northern Eurasia, reaching a peak of 1,300 to 1,500 animals in the early 20th century, and falling off as the numbers began dwindling.

In the first half of the 20th century, mechanized and overpoweringly efficient methods of hunting and trapping came into use in North America as well. Polar bears were chased from snowmobiles, icebreakers, and airplanes, the latter practice described in a 1965 *New York Times* editorial as being "about as sporting as machine gunning a cow." Norwegians used "self-killing guns", comprising a loaded rifle in a baited box that was placed at the level of a bear's head, and which fired when the string attached to the bait was pulled. The numbers taken grew rapidly in the 1960s, peaking around 1968 with a global total of 1,250 bears that year.

Contemporary Regulations

Road sign warning about the presence of polar bears

Concerns over the future survival of the species led to the development of national regulations on polar bear hunting, beginning in the mid-1950s. The Soviet Union banned all hunting in 1956. Canada began imposing hunting quotas in 1968. Norway passed a series of increasingly strict regulations from 1965 to 1973, and has completely banned hunting since then. The United States began regulating hunting in 1971 and adopted the Marine Mammal Protection Act in 1972. In 1973, the International Agreement on the Conservation of Polar Bears was signed by all five nations whose territory is inhabited by polar bears: Canada, Denmark, Norway, the Soviet Union, and the United States. Member countries agreed to place restrictions on recreational and commercial hunting, ban hunting from aircraft and icebreakers, and conduct further research. The treaty allows hunting "by local people using traditional methods". Norway is the only country of the five in which all harvest of polar bears is banned. The agreement was a rare case of international cooperation during the Cold War. Biologist Ian Stirling commented, "For many years, the conservation of polar bears was the only subject in the entire Arctic that nations from both sides of the Iron Curtain could agree upon sufficiently to sign an agreement. Such was the intensity of human fascination with this magnificent predator, the only marine bear."

Agreements have been made between countries to co-manage their shared polar bear subpopulations. After several years of negotiations, Russia and the United States signed an agreement in October 2000 to jointly set quotas for indigenous subsistence hunting in Alaska and Chukotka. The treaty was ratified in October 2007. In September 2015, the polar bear range states agreed upon a "circumpolar action plan" describing their conservation strategy for polar bears.

Although the United States government has proposed that polar bears be transferred to Appendix I of CITES, which would ban all international trade in polar bear parts, polar bears currently remain listed under Appendix II. This decision was approved of by members of the IUCN and TRAFFIC, who determined that such an uplisting was unlikely to confer a conservation benefit.

Canada

Dogsleds are used for recreational hunting of polar bears in Canada.

Polar bears were designated "Not at Risk" in April 1986 and uplisted to "Special Concern" in April 1991. This status was re-evaluated and confirmed in April 1999, November 2002, and April 2008. Polar bears continue to be listed as a species of special concern in Canada because of their sensitivity to overharvest and because of an expected range contraction caused by loss of Arctic sea ice.

More than 600 bears are killed per year by humans across Canada, a rate calculated by scientists to be unsustainable for some areas, notably Baffin Bay. Canada has allowed sport hunters accompanied by local guides and dog-sled teams since 1970, but the practice was not common until the 1980s. The guiding of sport hunters provides meaningful employment and an important source of income for northern communities in which economic opportunities are few. Sport hunting can bring CDN$20,000 to $35,000 per bear into northern communities, which until recently has been mostly from American hunters.

The territory of Nunavut accounts for the location 80% of annual kills in Canada. In 2005, the government of Nunavut increased the quota from 400 to 518 bears, despite protests from the IUCN Polar Bear Specialist Group. In two areas where harvest levels have been increased based on increased sightings, science-based studies have indicated declining populations, and a third area is considered data-deficient. While most of that quota is hunted by the indigenous Inuit people, a growing share is sold to recreational hunters. (0.8% in the 1970s, 7.1% in the 1980s, and 14.6% in the 1990s) Nunavut polar bear biologist, Mitchell Taylor, who was formerly responsible for polar bear conservation in the territory, has insisted that bear numbers are being sustained under current hunting limits. In 2010, the 2005 increase was partially reversed. Government of Nunavut officials announced that the polar bear quota for the Baffin Bay region would be gradually reduced from 105 per year to 65 by the year 2013. The Government of the Northwest Territories maintain their own quota of 72 to 103 bears within the Inuvialuit communities of which some are set aside for sports hunters. Environment Canada also banned the export from Canada of fur, claws, skulls and other products from polar bears harvested in Baffin Bay as of 1 January 2010.

Because of the way polar bear hunting quotas are managed in Canada, attempts to discourage sport hunting would actually increase the number of bears killed in the short term. Canada allocates a certain number of permits each year to sport and subsistence hunting, and those that are not used for sport hunting are re-allocated to indigenous subsistence hunting. Whereas northern communities kill all the polar bears they are permitted to take each year, only half of sport hunters with permits actually manage to kill a polar bear. If a sport hunter does not kill a polar bear before his or her permit expires, the permit cannot be transferred to another hunter.

In August 2011, Environment Canada published a national polar bear conservation strategy.

Greenland

In Greenland, hunting restrictions were first introduced in 1994 and expanded by executive order in 2005. Until 2005 Greenland placed no limit on hunting by indigenous people. However, in 2006 it imposed a limit of 150, while also allowed recreational hunting for the first time. Other provisions included year-round protection of cubs and mothers, restrictions on weapons used, and various administrative requirements to catalogue kills.

Russia

The Soviet Union banned the harvest of polar bears in 1956; however, poaching continued and is estimated to pose a serious threat to the polar bear population. In recent years, polar bears have approached coastal villages in Chukotka more frequently due to the shrinking of the sea ice, endangering humans and raising concerns that illegal hunting would become even more prevalent. In 2007, the Russian government made subsistence hunting legal for indigenous Chukotkan peoples only, a move supported by Russia's most prominent bear researchers and the World Wide Fund for Nature as a means to curb poaching.

Polar bears are currently listed as "Rare", of "Uncertain Status", or "Rehabilitated and rehabilitating" in the Red Data Book of Russia, depending on population. In 2010, the Ministry of Natural Resources and Environment published a strategy for polar bear conservation in Russia.

United States

The Marine Mammal Protection Act of 1972 afforded polar bears some protection in the United States. It banned hunting (except by indigenous subsistence hunters), banned importing of polar bear parts (except polar bear pelts taken legally in Canada), and banned the harassment of polar bears. On 15 May 2008, the United States Department of the Interior listed the polar bear as a threatened species under the Endangered Species Act, citing the melting of Arctic sea ice as the primary threat to the polar bear. It banned all importing of polar bear trophies. Importing products made from polar bears had been prohibited from 1972 to 1994 under the Marine Mammal Protection Act, and restricted between 1994 and 2008. Under those restrictions, permits from the United States Fish and Wildlife Service were required to import sport-hunted polar bear trophies taken in hunting expeditions in Canada. The permit process required that the bear be taken from an area with quotas based on sound management principles. Since 1994, hundreds of sport-hunted polar bear trophies have been imported into the U.S. In 2015, the U.S. Fish and Wildlife Service published a draft conservation management plan for polar bears to improve their status under the Endangered Species Act and the Marine Mammal Protection Act.

Conservation Status, Threats and Controversies

Polar bear population sizes and trends are difficult to estimate accurately because they occupy remote home ranges and exist at low population densities. Polar bear fieldwork can also be hazardous to researchers. As of 2015, the International Union for Conservation of Nature (IUCN) reports that the global population of polar bears is 22,000 to 31,000, and the current population trend is unknown. Nevertheless, polar bears are listed as "Vulnerable" under criterion A3c, which indicates an expected population decrease of ≥30% over the next three generations (~34.5 years) due to "decline in area of occupancy, extent of occurrence and/or quality of habitat". Risks to the polar bear include climate change, pollution in the form of toxic contaminants, conflicts with shipping, oil and gas exploration and development, and human-bear interactions including harvesting and possible stresses from recreational polar-bear watching.

According to the World Wildlife Fund, the polar bear is important as an indicator of Arctic ecosystem health. Polar bears are studied to gain understanding of what is happening throughout the Arctic, because at-risk polar bears are often a sign of something wrong with the Arctic marine ecosystem.

Climate Change

The International Union for Conservation of Nature, Arctic Climate Impact Assessment, United States Geological Survey and many leading polar bear biologists have expressed grave concerns about the impact of climate change, including the belief that the current warming trend imperils the survival of the polar bear.

Mothers and cubs have high nutritional requirements, which are not met if the seal-hunting season is too short

The key danger posed by climate change is malnutrition or starvation due to habitat loss. Polar bears hunt seals from a platform of sea ice. Rising temperatures cause the sea ice to melt earlier in the year, driving the bears to shore before they have built sufficient fat reserves to survive the period of scarce food in the late summer and early fall. Reduction in sea-ice cover also forces bears to swim longer distances, which further depletes their energy stores and occasionally leads to drowning. Thinner sea ice tends to deform more easily, which appears to make it more difficult for polar bears to access seals. Insufficient nourishment leads to lower reproductive rates in adult females and lower survival rates in cubs and juvenile bears, in addition to poorer body condition in bears of all ages.

In addition to creating nutritional stress, a warming climate is expected to affect various other aspects of polar bear life: Changes in sea ice affect the ability of pregnant females to build suitable maternity dens. As the distance increases between the pack ice and the coast, females must swim longer distances to reach favored denning areas on land. Thawing of permafrost would affect the bears who traditionally den underground, and warm winters could result in den roofs collapsing or having reduced insulative value. For the polar bears that currently den on multi-year ice, increased ice mobility may result in longer distances for mothers and young cubs to walk when they return to seal-hunting areas in the spring. Disease-causing bacteria and parasites would flourish more readily in a warmer climate.

Problematic interactions between polar bears and humans, such as foraging by bears in garbage dumps, have historically been more prevalent in years when ice-floe breakup occurred early and local polar bears were relatively thin. Increased human-bear interactions, including fatal attacks on humans, are likely to increase as the sea ice shrinks and hungry bears try to find food on land.

Polar bear on Svalbard, starving due to the ice around the islands melting earlier than before

The effects of climate change are most profound in the southern part of the polar bear's range, and this is indeed where significant degradation of local populations has been observed. The Western Hudson Bay subpopulation, in a southern part of the range, also happens to be one of the best-studied polar bear subpopulations. This subpopulation feeds heavily on ringed seals in late spring, when newly weaned and easily hunted seal pups are abundant. The late spring hunting season ends for polar bears when the ice begins to melt and break up, and they fast or eat little during the summer until the sea freezes again.

Due to warming air temperatures, ice-floe breakup in western Hudson Bay is currently occurring three weeks earlier than it did 30 years ago, reducing the duration of the polar bear feeding season. The body condition of polar bears has declined during this period; the average weight of lone (and likely pregnant) female polar bears was approximately 290 kg (640 lb) in 1980 and 230 kg (510 lb) in 2004. Between 1987 and 2004, the Western Hudson Bay population declined by 22%, although the population is currently listed as "stable". As the climate change melts sea ice, the U.S. Geological Survey projects that two-thirds of polar bears will disappear by 2050

In Alaska, the effects of sea ice shrinkage have contributed to higher mortality rates in polar bear cubs, and have led to changes in the denning locations of pregnant females. In recent years, polar bears in the Arctic have undertaken longer than usual swims to find prey, possibly resulting in four recorded drownings in the unusually large ice pack regression of 2005.

A new development is that polar bears have begun ranging to new territory. While not unheard of but still uncommon, polar bears have been sighted increasingly in larger numbers ashore, staying on the mainland for longer periods of time during the summer months, particularly in North Canada, traveling farther inland. This may cause an increased reliance on terrestrial diets, such as goose eggs, waterfowl and caribou, as well as increased human–bear conflict.

Pollution

Polar bears accumulate high levels of persistent organic pollutants such as polychlorinated biphenyl (PCBs) and chlorinated pesticides. Due to their position at the top of the ecological pyramid, with a diet heavy in blubber in which halocarbons concentrate, their bodies are among the most contaminated of Arctic mammals. Halocarbons are known to be toxic to other animals, because they mimic hormone chemistry, and biomarkers such as immunoglobulin G and retinol suggest similar effects on polar bears. PCBs have received the most study, and they have been associated with birth defects and immune system deficiency.

Many chemicals, such as PCBs and DDT, have been internationally banned due to the recognition of their harm on the environment. Their concentrations in polar bear tissues continued to rise for decades after being banned as these chemicals spread through the food chain. Since then, the trend seems to have discontinued, with tissue concentrations of PCBs declining between studies performed from 1989 to 1993 and studies performed from 1996 to 2002. During the same time periods, DDT was notably lower in the Western Hudson Bay population only.

Oil and Gas Development

Oil and gas development in polar bear habitat can affect the bears in a variety of ways. An oil spill in the Arctic would most likely concentrate in the areas where polar bears and their prey are also concentrated, such as sea ice leads. Because polar bears rely partly on their fur for insulation and soiling of the fur by oil reduces its insulative value, oil spills put bears at risk of dying from hypothermia. Polar bears exposed to oil spill conditions have been observed to lick the oil from their fur, leading to fatal kidney failure. Maternity dens, used by pregnant females and by females with infants, can also be disturbed by nearby oil exploration and development. Disturbance of these sensitive sites may trigger the mother to abandon her den prematurely, or abandon her litter altogether.

Predictions

Steven Amstrup and other U.S. Geological Survey scientists have predicted two-thirds of the world's polar bears may disappear by 2050, based on moderate projections for the shrinking of summer sea ice caused by climate change, though the validity of this study has been debated. The bears could disappear from Europe, Asia, and Alaska, and be depleted from the Canadian Arctic Archipelago and areas off the northern Greenland coast. By 2080, they could disappear from Greenland entirely and from the northern Canadian coast, leaving only dwindling numbers in the interior Arctic Archipelago. However, in the short term, some polar bear populations in historically colder regions of the Arctic may temporarily benefit from a milder climate, as multiyear ice that is too thick for seals to create breathing holes is replaced by thinner annual ice.

Polar bears diverged from brown bears 400,000–600,000 years ago and have survived past periods of climate fluctuation. It has been claimed that polar bears will be able to adapt to terrestrial food sources as the sea ice they use to hunt seals disappears. However, most polar bear biologists think that polar bears will be unable to completely offset the loss of calorie-rich seal blubber with terrestrial foods, and that they will be outcompeted by brown bears in this terrestrial niche, ultimately leading to a population decline.

Controversy Over Species Protection

Swimming

Warnings about the future of the polar bear are often contrasted with the fact that worldwide population estimates have increased over the past 50 years and are relatively stable today. Some estimates of the global population are around 5,000 to 10,000 in the early 1970s; other estimates were 20,000 to 40,000 during the 1980s. Current estimates put the global population at between 20,000 and 25,000 or 22,000 and 31,000.

There are several reasons for the apparent discordance between past and projected population trends: estimates from the 1950s and 1960s were based on stories from explorers and hunters rather than on scientific surveys. Second, controls of harvesting were introduced that allowed this previously overhunted species to recover. Third, the recent effects of climate change have affected sea ice abundance in different areas to varying degrees.

Debate over the listing of the polar bear under endangered species legislation has put conservation groups and Canada's Inuit at opposing positions; the Nunavut government and many northern residents have condemned the U.S. initiative to list the polar bear under the Endangered Species Act. Many Inuit believe the polar bear population is increasing, and restrictions on commercial sport-hunting are likely to lead to a loss of income to their communities.

In Culture

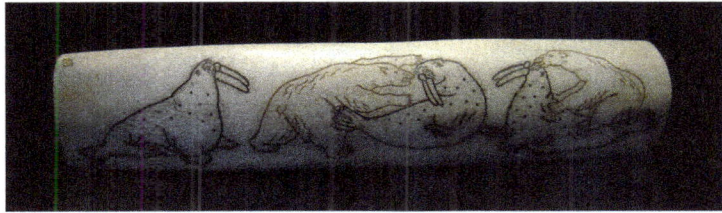

Engraving, made by Chukchi carvers in the 1940s on a walrus tusk, depicts polar bears hunting walrus

Indigenous Folklore

For the indigenous peoples of the Arctic, polar bears have long played an important cultural and material role. Polar bear remains have been found at hunting sites dating to 2,500 to 3,000 years ago and 1,500-year-old cave paintings of polar bears have been found in the Chukchi Peninsula. Indeed, it has been suggested that Arctic peoples' skills in seal hunting and igloo construction has been in part acquired from the polar bears themselves.

The Inuit and Alaska Natives have many folk tales featuring the bears including legends in which bears are humans when inside their own houses and put on bear hides when going outside, and stories of how the constellation that is said to resemble a great bear surrounded by dogs came into being. These legends reveal a deep respect for the polar bear, which is portrayed as both spiritually powerful and closely akin to humans. The human-like posture of bears when standing and sitting, and the resemblance of a skinned bear carcass to the human body, have probably contributed to the belief that the spirits of humans and bears were interchangeable. Eskimo legends tell of humans learning to hunt from the polar bear.

Among the Chukchi and Yupik of eastern Siberia, there was a longstanding shamanistic ritual of "thanksgiving" to the hunted polar bear. After killing the animal, its head and skin were removed and cleaned and brought into the home, a feast was held in the hunting camp in its honor. In order to appease the spirit of the bear, there were traditional song and drum music and the skull would be ceremonially fed and offered a pipe. Only once the spirit was appeased would the skull be separated from the skin, taken beyond the bounds of the homestead, and placed in the ground, facing north. Many of these traditions have faded somewhat in time, especially in light of the total hunting ban in the Soviet Union (and now Russia) since 1955.

The Nenets of north-central Siberia placed particular value on the talismanic power of the prominent canine teeth. They were traded in the villages of the lower Yenisei and Khatanga rivers to the forest-dwelling peoples further south, who would sew them into their hats as protection against brown bears. It was believed that the "little nephew" (the brown bear) would not dare to attack a man wearing the tooth of its powerful "big uncle" (the polar bear). The skulls of killed polar bears were buried at specific sacred sites and altars, called *sedyangi*, were constructed out of the skulls. Several such sites have been preserved on the Yamal Peninsula.

Symbols and Mascots

Their distinctive appearance and their association with the Arctic have made polar bears popular icons, especially in those areas where they are native. The Canadian Toonie (two-dollar coin) features the image of a polar bear and both the Northwest Territories and Nunavut license plates in Canada are in the shape of a polar bear. The polar bear is the mascot of Bowdoin College in Maine and the University of Alaska Fairbanks and was chosen as mascot for the 1988 Winter Olympics held in Calgary. The Eisbären Berlin professional hockey team, playing in the DEL top-level pro hockey league of Germany uses a roaring polar bear (seen head-on) as their team logo.

Greenland's 1911 five kroner note depicting a polar bear

Coat of arms of the Chukotka Autonomous Okrug in the Russian Federation

Coat of arms of the Greenlandic Self-Rule government (Kalaallit Nunaat)

Companies such as Coca-Cola, Polar Beverages, Nelvana, Bundaberg Rum, and Good Humor-Breyers have used images of the polar bear in advertising, while Fox's Glacier Mints have featured a polar bear named Peppy as the brand mascot since 1922. This has supported the popularity of the polar bear, and it has since become one of a collection of creatures who are associated with Christmas, including penguins, reindeer and the European robin.

Fiction

Polar bears are also popular in fiction, particularly in books aimed at children or teenagers. For example, *The Polar Bear Son* is adapted from a traditional Inuit tale. The animated television series *Noah's Island* features a polar bear named Noah as the protagonist. Polar bears feature prominently in *East* (also released as *North Child*) by Edith Pattou, *The Bear* by Raymond Briggs (adapted into an animated short in 1998), and Chris d'Lacey's *The Fire Within* series. The *panserbjørne* of

Philip Pullman's fantasy trilogy *His Dark Materials* are sapient, dignified polar bears who exhibit anthropomorphic qualities, and feature prominently in the 2007 film adaptation of *The Golden Compass*. The television series *Lost* features polar bears living on the tropical island setting.

Cetacea

Cetacea are a widely distributed and diverse clade of carnivorous, finned, aquatic marine mammals. They comprise the extant parvorders Odontoceti (toothed whales including dolphins and porpoises), Mysticeti (the baleen whales), and Archaeoceti (the ancestors of modern whales, and now extinct). There are around 89 species of cetaceans, and more than 70 belonging to Odontoceti. While cetaceans were historically thought to have descended from mesonychids, molecular evidence supports them as a relative of Artiodactyls (even-toed ungulates). Cetaceans belong to the order Cetartiodactyla (formed by combining Cetacea + Artiodactyla) and their closest living relatives are hippopotamuses and other hoofed mammals (camels, pigs, and ruminants), having diverged about 50 million years ago.

Cetaceans range in size from the 1 m (3 ft 3 in) and 50 kg (110 lb) Maui's dolphin to the 29.9 m (98 ft) and 190,000 kg (420,000 lb) blue whale, which is also the largest animal ever known to have existed. Several species exhibit sexual dimorphism. They have streamlined bodies and two (external) limbs that are modified into flippers. Though not as flexible or agile as seals, cetaceans can swim very fast, with the killer whale able to travel at 56 kilometres per hour (35 mph) in short bursts and the fin whale able to cruise at 48 kilometres per hour (30 mph). The hindlimbs of cetaceans are internal, and are thought to be vestigial. Dolphins are able to make very tight turns while swimming at high speeds. Baleen whales have short hairs on their mouth, unlike the toothed whales. Cetaceans have well-developed senses—their eyesight and hearing are adapted for both air and water, and baleen whales have a tactile system in their vibrissae. Some species are well adapted for diving to great depths. They have a layer of fat, or blubber, under the skin to keep warm in the cold water.

Although cetaceans are widespread, most species prefer the colder waters of the Northern and Southern Hemispheres. They spend their lives in the water, having to mate, give birth, molt or escape from predators, like killer whales, underwater. This has drastically affected their anatomy to be able to do so. They feed largely on fish and marine invertebrates; but a few, like the killer whale, feed on large mammals and birds, such as penguins and seals. Some baleen whales (mainly gray whales and right whales) are specialised for feeding on benthic creatures. Male cetaceans typically mate with more than one female (polygyny), although the degree of polygyny varies with the species. Cetaceans are not shown to have pair bonds. Male cetacean strategies for reproductive success vary between herding females, defending potential mates from other males, or whale song which attracts mates. Calves are typically born in the fall and winter months, and females bear almost all the responsibility for raising them. Mothers of some species fast and nurse their young for a relatively short period of time, which is more typical of baleen whales as their main food source (invertebrates) aren't found in their breeding and calving grounds (tropics). Cetaceans produce a number of vocalizations, notably the clicks and whistles of dolphins, the moaning songs of the humpback whale.

The meat, blubber and oil of cetaceans have traditionally been used by indigenous peoples of the Arctic. Cetaceans have been depicted in various cultures worldwide. Dolphins are commonly kept in captivity and are even sometimes trained to perform tricks and tasks, other cetaceans aren't as often kept in captivity (with usually unsuccessful attempts). Once relentlessly hunted by commercial industries for their products, cetaceans are now protected by international law. The baiji (Chinese river dolphin) has become "Possibly Extinct" in the past century, while the vaquita and Yangtze finless porpoise are ranked Critically Endangered by the International Union for Conservation of Nature. Besides hunting, cetaceans also face threats from accidental trapping, marine pollution, and ongoing climate change.

Baleen Whales and Toothed Whales

The two parvorders, baleen whales (Mysticeti) and toothed whales (Odontoceti), are thought to have diverged around thirty-four million years ago.

Baleen whales have bristles made of keratin instead of teeth. The bristles filter krill and other small invertebrates from seawater. Grey whales feed on bottom-dwelling mollusks. Rorqual family (balaenopterids) use throat pleats to expand their mouths to take in food and sieve out the water. Balaenids (right whales and bowhead whales) have massive heads that can make up 40% of their body mass. Most mysticetes prefer the food-rich colder waters of the Northern and Southern Hemispheres, migrating to the Equator to give birth. During this process, they are capable of fasting for several months, relying on their fat reserves.

The parvorder of Odontocetes – the toothed whales – include sperm whales, beaked whales, killer whales, dolphins and porpoises. They have conical teeth designed for catching fish or squid. A few, such as the killer whale, feed on mammals, such as pinnipeds and other whales. They have well-developed senses – their eyesight and hearing are adapted for both air and water, and they have advanced sonar capabilities using their melon. Their hearing is so well-adapted for both air and water that some blind specimens can survive. Some species, such as sperm whales, are well-adapted for diving to great depths. Several species of odontocetes show sexual dimorphism, in which the males differ from the females, usually for purposes of sexual display or aggression. Toothed whales feed largely on fish and marine invertebrates.

Anatomy

Cetacean bodies are generally similar to that of fish, which can be attributed to their lifestyle and the habitat conditions. Their body is well-adapted to their habitat, although they share essential characteristics with other higher mammals (Eutheria):

They have a streamlined shape, and their forelimbs are flippers. Almost all have a dorsal fin on their backs that can take on many forms depending on the species. A few species, such as the beluga whale, lack them. Both the flipper and the fin are for stabilization and steering in the water.

The male genitals and mammary glands of females are sunken into the body.

The body is wrapped in a thick layer of fat, known as blubber, used for thermal insulation and gives cetaceans their smooth, streamlined body shape. In larger species, it can reach a thickness up to half a meter (1.6 ft).

Sexual dimorphism evolved in many toothed whales. Sperm whales, narwhals, many members of the beaked whale family, several species of the porpoise family, killer whales, pilot whales, eastern spinner dolphins and northern right whale dolphins show this characteristic. Males in these species developed external features absent in females that are advantageous in combat or display. For example, male sperm whales are up to 63% percent larger than females, and many beaked whales possess tusks used in competition among males.

Fluke

Humpback whale fluke

They have a cartilaginous fluke at the end of their tails that is used for propulsion. The fluke is set horizontally on the body, unlike fish, which have vertical tails.

Hind legs are not present in cetaceans, nor are any other external body attachments such as a pinna and hair.

Head

Whales have an elongated head, especially baleen whales, due to the wide overhanging jaw. Bowhead whale plates can be 4 metres (13 ft) long. Their nostril(s) make up the blowhole, with one in toothed whales and two in baleen whales.

The nostrils are located on top of the head above the eyes so that the rest of the body can remain submerged while surfacing for air. The back of the skull is significantly shortened and deformed. By shifting the nostrils to the top of the head, the nasal passages extend perpendicularly through the skull. The teeth or baleen in the upper jaw sit exclusively on the maxilla. The braincase is concentrated through the nasal passage to the front and is correspondingly higher, with individual cranial bones that overlap. The bony otic capsule, the petrosal, is connected to the skull with cartilage, so that it can swing independently.

In toothed whales, connective tissue exists in the melon as a head buckle. This is filled with air sacs and fat that aid in buoyancy and biosonar. The sperm whale has a particularly pronounced melon; this is called the spermaceti organ and contains the eponymous spermaceti, hence the name "sperm whale". Even the long tusk of the narwhal is a vice-formed tooth. In many toothed whales, the depression in their skull is due to the formation of a large melon and multiple, asymmetric air bags.

River dolphins, unlike most other cetaceans, can turn their head 90°. Other cetaceans have fused neck vertebrae and are unable to turn their head at all.

The baleen of baleen whales consists of long, fibrous strands of keratin. Located in place of the teeth, it has the appearance of a huge fringe and is used to sieve the water for plankton and krill.

Brain

The neocortex of many cetaceans is home to elongated spindle neurons that, prior to 2007, were known only in hominids. In humans, these cells are involved in social conduct, emotions, judgment and theory of mind. Cetacean spindle neurons are found in areas of the brain homologous to where they are found in humans, suggesting they perform a similar function.

Brain size was previously considered a major indicator of intelligence. Since most of the brain is used for maintaining bodily functions, greater ratios of brain to body mass may increase the amount of brain mass available for cognitive tasks. Allometric analysis indicates that mammalian brain size scales at approximately two-thirds or three-quarter exponent of the body mass. Comparison of a particular animal's brain size with the expected brain size based on such an analysis provides an encephalization quotient that can be used as an indication of animal intelligence. Sperm whales have the largest brain mass of any animal on earth, averaging 8,000 cm³ (490 in³) and 7.8 kg (17 lb) in mature males. The brain to body mass ratio in some odontocetes, such as belugas and narwhals, is second only to humans. In some whales, however, it is less than half that of humans: 0.9% versus 2.1%. The sperm whale (*Physeter macrocephalus*) is the largest of all toothed predatory animals and possesses the largest brain.

Skeleton

Cetacea Skeletons

The cetacean skeleton is largely made up of cortical bone, which stabilizes the animal in the water. For this reason, the usual terrestrial compact bones, which are finely woven cancellous bone, are replaced with lighter and more elastic material. In many places, bone elements are replaced by cartilage and even fat, thereby improving their hydrostatic qualities. The ear and the muzzle contain a bone shape that is exclusive to cetaceans with a high density, resembling porcelain. This conducts sound better than other bones, thus aiding biosonar.

Skeleton of a blue whale standing outside the Long Marine Laboratory of the University of California, Santa Cruz.

Weathered upper jaw of a sperm whale.

Bowhead whale skeleton

Sperm whale skeleton

The number of vertebrae that make up the spine varies by species, ranging from forty to ninety-three. The cervical spine, found in all mammals, consists of seven vertebrae which, however, are reduced or fused. This gives stability during swimming at the expense of mobility. The fins are carried by the thoracic vertebrae, ranging from nine to seventeen individual vertebrae. The sternum is cartilaginous. The last two to three pairs of ribs are not connected and hang freely in the body wall. The stable lumbar and tail include the other vertebrae. Below the caudal vertebrae is the chevron bone; the vortex developed provides additional attachment points for the tail musculature.

The front limbs are paddle-shaped with shortened arms and elongated finger bones, to support movement. They are connected by cartilage. The second and third fingers display a proliferation of the finger members, a so-called hyperphalangy. The shoulder joint is the only functional joint in all cetaceans except for the Amazon river dolphin. The collarbone is completely absent. The rear limbs are vestigial, without connections to the spine.

Physiology

Circulation

Cetaceans have powerful hearts. Blood oxygen is distributed effectively throughout the body. They are warm-blooded, i.e., they hold a nearly constant body temperature.

Respiration

Cetaceans have lungs, meaning they breathe air. An individual can last without a breath from a few minutes to over two hours depending on the species. Cetacea are deliberate breathers who must be awake to inhale and exhale. When stale air, warmed from the lungs, is exhaled, it condenses as it meets colder external air. As with a terrestrial mammal breathing out on a cold day, a small cloud of 'steam' appears. This is called the 'spout' and varies across species in shape, angle and height. Species can be identified at a distance using this characteristic.

The structure of the respiratory and circulatory systems is of particular importance for the life of marine mammals. The oxygen balance is effective. Each breath can replace up to 90% of the total lung volume. For land mammals, in comparison, this value is usually about 15%. During inhalation, about twice as much oxygen is absorbed by the lung tissue as in a land mammal. As with all mammals, the oxygen is stored in the blood and the lungs, but in cetaceans, it is also stored in various tissues, mainly in the muscles. The muscle pigment, myoglobin, provides an effective bond.

This additional oxygen storage is vital for deep diving, since beyond a depth around 100 m (330 ft), the lung tissue is almost completely compressed by the water pressure.

Organs

The stomach consists of three chambers. The first region is formed by a loose gland and a muscular forestomach (missing in beaked whales), which is then followed by the main stomach and the pylorus. Both are equipped with glands to help digestion. A bowel adjoins the stomachs, whose individual sections can only be distinguished histologically. The liver is large and separate from the gall bladder.

The kidneys are long and flattened. The salt concentration in cetacean blood is lower than that in seawater, requiring kidneys to excrete salt. This allows the animals to drink seawater.

Senses

Cetacean eyes are set on the sides rather than the front of the head. This means only species with pointed 'beaks' (such as dolphins) have good binocular vision forward and downward. Tear glands secrete greasy tears, which protect the eyes from the salt in the water. The lens is almost spherical, which is most efficient at focusing the minimal light that reaches deep water. Cetaceans make up for their generally poor vision (except dolphins) with excellent hearing.

At least one species, the tucuxi or Guiana dolphin, is able to use electroreception to sense prey.

Teeth/Baleen

While the teeth are divided into incisors, canines and molars among terrestrial archaeocetes, the teeth of modern cetaceans are brought into line with each other, which can be seen among the fish-eating odontocetes (transition from heterodont to homodont).

Ears

Biosonar

The external ear has lost the pinna (visible ear), but still retains a narrow external auditory meatus. To register sounds, instead, the posterior part of the mandible has a thin lateral wall (the pan bone) fronting a concavity that houses a fat pad. The pad passes anteriorly into the greatly enlarged mandibular foramen to reach in under the teeth and posteriorly to reach the thin lateral wall of the ectotympanic. The ectotympanic offers a reduced attachment area for the tympanic membrane. The connection between this auditory complex and the rest of the skull is reduced—to a single, small cartilage in oceanic dolphins.

In odontocetes, the complex is surrounded by spongy tissue filled with air spaces, while in mysticetes, it is integrated into the skull as with land mammals. In odontocetes, the tympanic membrane (or ligament) has the shape of a folded-in umbrella that stretches from the ectotympanic ring and narrows off to the malleus (quite unlike the flat, circular membrane found in land mammals.) In mysticetes, it also forms a large protrusion (known as the "glove finger"), which stretches into the external meatus and the stapes are larger than in odontocetes. In some small sperm whales, the malleus is fused with the ectotympanic.

The ear ossicles are pachyosteosclerotic (dense and compact) and differently shaped from land mammals (other aquatic mammals, such as sirenians and earless seals, have also lost their pinnae). T semicircular canals are much smaller relative to body size than in other mammals.

The auditory bulla is separated from the skull and composed of two compact and dense bones (the periotic and tympanic) referred to as the tympanoperiotic complex. This complex is located in a cavity in the middle ear, which, in the Mysticeti, is divided by a bony projection and compressed between the exoccipital and squamosal, but in the odontoceti, is large and completely surrounds the bulla (hence called "peribullar"), which is, therefore, not connected to the skull except in physeterids. In the Odontoceti, the cavity is filled with a dense foam in which the bulla hangs suspended in five or more sets of ligaments. The pterygoid and peribullar sinuses that form the cavity tend to be more developed in shallow water and riverine species than in pelagic Mysticeti. In Odontoceti, the composite auditory structure is thought to serve as an acoustic isolator, analogous to the lamellar construction found in the temporal bone in bats.

Cetaceans use sound to communicate, using groans, moans, whistles, clicks or the 'singing' of the humpback whale.

Echolocation

Odontoceti are generally capable of echolocation. They can discern the size, shape, surface characteristics, distance and movement of an object. They can search for, chase and catch fast-swimming prey in total darkness. Most Odontoceti can distinguish between prey and nonprey (such as humans or boats); captive Odontoceti can be trained to distinguish between, for example, balls of different sizes or shapes.

Mysticeti have exceptionally thin, wide basilar membranes in their cochleae without stiffening agents, making their ears adapted for processing low to infrasonic frequencies. Echolocation clicks also contain characteristic details unique to each animal, which may suggest that toothed whales can discern between their own click and that of others.

Chromosomes

The initial karyotype includes a set of chromosomes from 2n = 44. They have four pairs of telocentric chromosomes (whose centromeres sit at one of the telomeres), two to four pairs of subtelocentric and one or two large pairs of submetacentric chromosomes. The remaining chromosomes are metacentric—the centromere is approximately in the middle—and are rather small. Sperm whales, beaked whales and right whales converge to a reduction in the number of chromosomes to 2n = 42.

Ecology

Range and Habitat

Cetaceans are found in all oceans. River dolphin species live exclusively in fresh water. While many marine species, such as the blue whale, the humpback whale and the killer whale, have a distribution area that includes nearly the entire ocean, some species occur only locally or in broken populations. These include the vacquita, which inhabits a small part of the Gulf of California and Hector's dolphin, which lives in some coastal waters in New Zealand. Both species prefer deeper marine areas and species that live frequently or exclusively in coastal and shallow water areas.

Many species inhabit specific latitudes, often in tropical or subtropical waters, such as Bryde's whale or Risso's dolphin. Others are found only in a specific body of water. The southern right whale dolphin and the hourglass dolphin live only in the Southern Ocean. The narwhal and the beluga live only in the Arctic Ocean. Sowerby's beaked whale and the Clymene dolphin exist only in the Atlantic and the Pacific white-sided dolphin and the northern straight dolphin live only in the North Pacific.

Cosmopolitan species may be found in the Pacific, Atlantic and Indian Oceans. However, northern and southern populations become genetically separated over time. In some species, this separation leads eventually to a divergence of the species, such as produced the southern right whale, North Pacific right whale and North Atlantic right whale. Migratory species' reproductive sites often lie in the tropics and their feeding grounds in polar regions.

Thirty-two species are found in European waters, including twenty-five toothed and seven baleen species.

Behaviour

Sleep

Conscious breathing cetaceans sleep but cannot afford to be unconscious for long, because they may drown. While knowledge of sleep in wild cetaceans is limited, toothed cetaceans in captivity have been recorded to exhibit unihemispheric slow-wave sleep (USWS), which means they sleep with one side of their brain at a time, so that they may swim, breathe consciously and avoid both predators and social contact during their period of rest.

A 2008 study found that sperm whales sleep in vertical postures just under the surface in passive shallow 'drift-dives', generally during the day, during which whales do not respond to passing vessels unless they are in contact, leading to the suggestion that whales possibly sleep during such dives.

Diving

While diving, the animals reduce their oxygen consumption by lowering the heart activity and blood circulation; individual organs receive no oxygen during this time. Some rorquals can dive for up to 40 minutes, sperm whales between 60 and 90 minutes and bottlenose whales for two hours. Diving depths average about 100 m (330 ft). Species such as sperm whales can dive to 3,000 m (9,800 ft), although more commonly 1,200 metres (3,900 ft).

Social Relations

Most whales are social animals, although a few species live in pairs or are solitary. A group, known as a pod, usually consists of ten to fifty animals, but on occasion, such as mass availability of food or during mating season, groups may encompass more than one thousand individuals. Inter-species socialization can occur.

Pods have a fixed hierarchy, with the priority positions determined by biting, pushing or ramming. The behavior in the group is aggressive only in situations of stress such as lack of food, but usually it is peaceful. Contact swimming, mutual fondling and nudging are common. The playful behavior of the animals, which is manifested in air jumps, somersaults, surfing, or fin hitting, occurs more often than not in smaller cetaceans, such as dolphins and porpoises.

Whale Song

Males in some baleen species communicate via whale song, sequences of high pitched sounds. These "songs" can be heard for hundreds of kilometers. Each population generally shares a distinct song, which evolves over time. Sometimes, an individual can be identified by its distinctive vocals, such as the 52-hertz whale that sings at a higher frequency than other whales. Some individuals are capable of generating over 600 distinct sounds. In baleen species such as humpbacks, blues and fins, male-specific song is believed to be used to attract and display fitness to females.

Hunting

Pod groups also hunt, often with other species. Many species of dolphins hunt accompany large tunas on hunting expeditions, following large schools of fish. The killer whale hunts in pods and targets belugas and even larger whales. Humpback whales, among others, form in collaboration bubble carpets to herd krill or plankton into bait balls before lunging at them.

Intelligence

Bubble net feeding

Cetacea are known to teach, learn, cooperate, scheme and grieve.

Smaller cetaceans, such as dolphins and porpoises, engage in complex play behavior, including such things as producing stable underwater toroidal air-core vortex rings or "bubble rings". The two main methods of bubble ring production are rapid puffing of air into the water and allowing it to rise to the surface, forming a ring, or swimming repeatedly in a circle and then stopping to inject air into the helical vortex currents thus formed. They also appear to enjoy biting the vortex rings, so that they burst into many separate bubbles and then rise quickly to the surface. Whales produce bubble nets to aid in herding prey.

Killer whale porpoising

Larger whales are also thought to engage in play. The southern right whale elevates its tail fluke above the water, remaining in the same position for a considerable time. This is known as "sailing". It appears to be a form of play and is most commonly seen off the coast of Argentina and South Africa. Humpback whales also display this behaviour.

Self-awareness appears to be a sign of abstract thinking. Self-awareness, although not well-defined, is believed to be a precursor to more advanced processes such as metacognitive reasoning (thinking about thinking) that humans exploit. Cetaceans appear to possess self-awareness. The most widely used test for self-awareness in animals is the mirror test, in which a temporary dye is placed on an animal's body and the animal is then presented with a mirror. Researchers then explore whether the animal shows signs of self-recognition.

Critics claim that the results of these tests are susceptible to the Clever Hans effect. This test is much less definitive than when used for primates. Primates can touch the mark or the mirror, while cetaceans cannot, making their alleged self-recognition behavior less certain. Skeptics argue that behaviors said to identify self-awareness resemble existing social behaviors, so researchers could be misinterpreting self-awareness for social responses. Advocates counter that the behaviors are different from normal responses to another individual. Cetaceans show less definitive behavior of self-awareness, because they have no pointing ability.

In 1995, Marten and Psarakos used video to test dolphin self-awareness. They showed dolphins real-time footage of themselves, recorded footage and another dolphin. They concluded that their evidence suggested self-awareness rather than social behavior. While this particular study has not been replicated, dolphins later "passed" the mirror test.

Life History

Reproduction and Brooding

Most cetaceans sexually mature at seven to 10 years. An exception to this is the La Plata dolphin, which is sexually mature at two years, but lives only to about 20. The sperm whale reaches sexual maturity within about 20 years and a lifespan between 50 and 100 years.

For most species, reproduction is seasonal. Ovulation coincides with male fertility. This cycle is usually coupled with seasonal movements that can be observed in many species. Most toothed whales have no fixed bonds. In many species, females choose several partners during a season. Baleen whales are largely monogamous within each reproductive period.

Gestation ranges from 9 to 16 months. Duration is not necessarily a function of size. Porpoises and blue whales gestate for about 11 months. During gestation, the embryo is fed by a special nutritive tissue, the placenta.

Cetaceans usually bear one calf. In the case of twins, one usually dies, because the mother cannot produce sufficient milk for both. The fetus is positioned for a tail-first delivery, so that the risk of drowning during delivery is minimal. After birth, the mother carries the infant to the surface for its first breath. At birth they are about one-third of their adult length and tend to be independently active, comparable to terrestrial mammals.

Suckling

Like other placental mammals, cetaceans give birth to well-developed calves and nurse them with milk from their mammary glands. When suckling, the mother actively splashes milk into the mouth of the calf, using the muscles of her mammary glands, as the calf has no lips. This milk usually has a high fat content, ranging from 16 to 46%, causing the calf to increase rapidly in size and weight.

In many small cetaceans, suckling lasts for about four months. In large species, it lasts for over a year and involves a strong bond between mother and offspring.

The mother is solely responsible for brooding. In some species, so-called "aunts" occasionally suckle the young.

This reproductive strategy provides a few offspring that have a high survival rate.

Lifespan

Among cetaceans, whales are distinguished by an unusual longevity compared to other higher mammals. Some species, such as the bowhead whale (*Balaena mysticetus*), can reach over 200 years. Based on the annual rings of the bony otic capsule, the age of the oldest known specimen is a male determined to be 211 years at the time of death.

Death

Upon death, whale carcasses fall to the deep ocean and provide a substantial habitat for marine life. Evidence of whale falls in present-day and fossil records shows that deep-sea whale falls support a rich assemblage of creatures, with a global diversity of 407 species, comparable to other neritic biodiversity hotspots, such as cold seeps and hydrothermal vents.

Deterioration of whale carcasses happens through three stages. Initially, organisms such as sharks and hagfish scavenge the soft tissues at a rapid rate over a period of months and as long as two years. This is followed by the colonization of bones and surrounding sediments (which contain organic matter) by enrichment opportunists, such as crustaceans and polychaetes, throughout a period of years. Finally, sulfophilic bacteria reduce the bones releasing hydrogen sulfide enabling the growth of chemoautotrophic organisms, which in turn, support organisms such as mussels, clams, limpets and sea snails. This stage may last for decades and supports a rich assemblage of species, averaging 185 per site.

Disease

Brucellosis affects almost all mammals. It is distributed worldwide, while fishing and pollution have caused porpoise population density pockets, which risks further infection and disease spreading. *Brucella ceti*, most prevalent in dolphins, has been shown to cause chronic disease, increasing the chance of failed birth and miscarriages, male infertility, neurobrucellosis, cardiopathies, bone and skin lesions, strandings and death. Until 2008, no case had ever been reported in porpoises, but isolated populations have an increased risk and consequentially a high mortality rate.

Phylogenetics

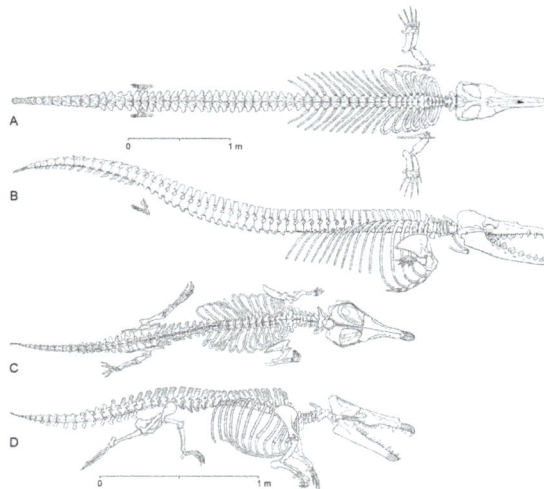

Comparison of the skeleton of *Dorudon atrox* and *Maiacetus inuus* in swimming position

Molecular biology and immunology show that cetaceans are phylogenetically closely related with the even-toed ungulates (Artiodactyla). Whales direct lineage began in the early Eocene, more than 50 million years ago, with early artiodactyls. Fossil discoveries at the beginning of the 21st century confirmed this.

Most molecular biological evidence suggests that hippos are the closest living relatives. Common anatomical features include similarities in the morphology of the posterior molars, and the bony ring on the temporal bone (bulla) and the involucre, a skull feature that was previously associated only with cetaceans. The fossil record, however, does not support this relationship, because the hippo lineage dates back only about 15 million years. The most striking common feature is the talus, a bone in the upper ankle. Early cetaceans, archaeocetes, show double castors, which only occur in even-toed ungulates. Corresponding findings are from Tethys Sea deposits in northern India and Pakistan. The Tethys Sea was a shallow sea between the Asian continent and northward-bound Indian plate.

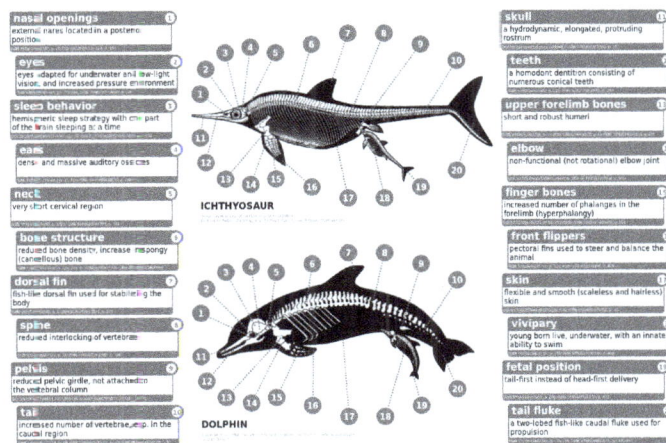

nasal openings
external nares located in a posterio position

eyes
eyes adapted for underwater and low-light vision, and increased pressure environment

sleep behavior
hemispheric sleep strategy with one part of the brain sleeping at a time

ears
dense and massive auditory ossicles

neck
very short cervical region

bone structure
reduced bone density, increase in spongy (cancellous) bone

dorsal fin
fish-like dorsal fin used for stabilizing the body

spine
reduced interlocking of vertebrae

pelvis
reduced pelvic girdle, not attached to the vertebral column

tail
increased number of vertebrae, esp. in the caudal region

ICHTHYOSAUR

DOLPHIN

skull
a hydrodynamic, elongated, protruding rostrum

teeth
a homodont dentition consisting of numerous conical teeth

upper forelimb bones
short and robust humeri

elbow
non-functional (not rotational) elbow joint

finger bones
increased number of phalanges in the forelimb (hyperphalangy)

front flippers
pectoral fins used to steer and balance the animal

skin
flexible and smooth (scaleless and hairless) skin

vivipary
young born live, underwater, with an innate ability to swim

fetal position
tail-first instead of head-first delivery

tail fluke
a two-lobed fish-like caudal fluke used for propulsion

Cetaceans display convergent evolution with fish and aquatic reptiles

Mysticetes evolved baleen around 25 million years ago and lost their teeth.

Development

Ancestors

The direct ancestors of today's cetaceans are probably found within the Dorudontidae whose most famous member, *Dorudon atrox*, lived at the same time as *Basilosaurus*. Both groups had already developed the typical anatomical features of today's whales, such as hearing. Life in the water for a formerly terrestrial creature required significant adjustments such as the fixed bulla, which replaces the mammalian eardrum, as well as sound-conducting elements for submerged directional hearing. Their wrists were stiffened and probably contributed to the typical build of flippers. The hind legs existed, however, but were significantly reduced in size and with a vestigial pelvis connection.

Transition from Land to Sea

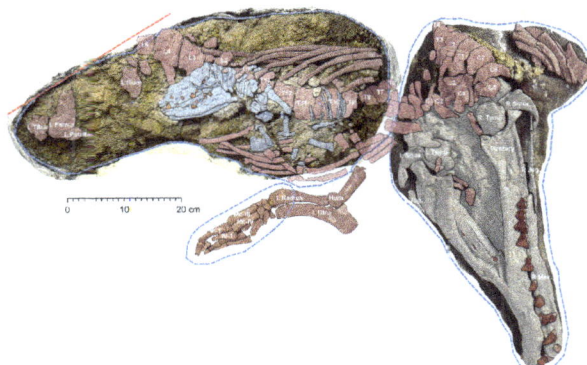

Fossil of a *Maiacetus* (red, beige skull) with fetus (blue, red teeth) shortly before the end of gestation

The fossil record traces the gradual transition from terrestrial to aquatic life. The regression of the hind limbs allowed greater flexibility of the spine. This made it possible for whales to move around with the vertical tail hitting the water. The front legs transformed into flippers, costing them their mobility on land.

One of the oldest members of ancient cetaceans (Archaeoceti) is *Pakicetus* from the Middle Eocene. This is an animal the size of a wolf, whose skeleton is known only partially. It had functioning legs and lived near the shore. This suggests the animal could still move on land. The long snout had carnivorous dentition.

The transition from land to sea dates to about 49 million years ago, with the *Ambulocetus* ("running whale"), discovered in Pakistan. It was up to 3 m (9.8 ft) long. The limbs of this archaeocete were adapted to swimming, but terrestrial locomotion was still possible. It probably crawled like a seal or crocodile. The snout was elongated with overhead nostrils and eyes. The tail was strong and supported movement through water. *Ambulocetus* probably lived in mangroves in brackish water and fed in the riparian zone as a predator of fish and other vertebrates.

Dating from about 45 million years ago are species such as *Indocetus, Kutchicetus, Rodhocetus* and *Andrewsiphius,* all of which were adapted to life in water. The hind limbs of these species were regressed and their body shapes resemble modern whales. Protocetidae family member *Rodhocetus* is considered the first to be fully aquatic. The body was streamlined and delicate with extended hand and foot bones. The merged pelvic lumbar spine was present, making it possible to support the floating movement of the tail. It was likely a good swimmer, but could probably move only clumsily on land, much like a modern seal.

Marine Animals

Since the late Eocene, about 40 million years ago, cetaceans populated the subtropical oceans and no longer emerged on land. An example is the 18-m-long *Basilosaurus,* sometimes referred to as *Zeuglodon.* The transition from land to water was completed in about 10 million years. The Wadi Al-Hitan ("Whale Valley") in Egypt contains numerous skeletons of *Basilosaurus,* as well as other marine vertebrates.

Taxonomy

Baleen whales (Mysticeti) owe their name to their baleen. Toothed whales (Odontoceti), which include the dolphins and porpoises, have conical teeth or spade-shaped teeth and can perceive their environment through biosonar.

The infraorder comprises the families Balaenidae (right and bowhead whales), Balaenoptera (rorquals), Eschrichtiidae (the gray whale), Delphinidae (oceanic dolphins), Monodontidae (Arctic whales), Phocoenidae (porpoises), Physeteridae (sperm whales), Kogiidae (lesser sperm whales), Platanistidae (Old World river dolphins), Iniidae (New World river dolphins), Pontoporiidae (the La plata dolphin) and Ziphidae (beaked whales).

- INFRAORDER CETACEA
  - Parvoder Mysticeti: Baleen whales
    - Superfamily Balaenoidea: Right whales
      - Family Balaenidae
        - Genus *Balaena*

- Bowhead whale, *Balaena mysticetus*
- Genus *Eubalaena*
  - North Atlantic right whale, *Eubalaena glacialis*
  - North Pacific right whale, *Eubalaena japonica*
  - Southern right whale, *Eubalaena australis*
- Family Cetotheriidae
  - Genus *Caperea*
    - Pygmy right whale, *Caperea marginata*
- Superfamily Balaenopteroidea
  - Family Balaenopteridae: Rorquals
    - Subfamily Balaenopterinae
      - Genus *Balaenoptera*: slender rorquals
        - Common minke whale, *Balaenoptera acuto-rostrata*
        - Antarctic minke whale, *Balaenoptera bonae-rensis*
        - Sei whale, *Balaenoptera borealis*
        - Bryde's whale, *Balaenoptera brydei*
        - Eden's whale *Balaenoptera edeni*
        - Omura's whale – *Balaenoptera omurai*
        - Blue whale, *Balaenoptera musculus*
        - Fin whale, *Balaenoptera physalus*
    - Subfamily Megapterinae
      - Genus *Megaptera*
        - Humpback whale, *Megaptera novaeangliae*
  - Family Eschrichtiidae: Gray whales
    - Genus *Eschrichtius*
      - Gray whale, *Eschrichtius robustus*
- Parvorder Odontoceti: Toothed whales
  - Superfamily Delphinoidea: Oceanic dolphins
    - Family Delphinidae

- Genus *Cephalorhynchus*: blunt-nosed dolphins
  - Commerson's dolphin, *Cephalorhyncus commersonii*
  - Chilean dolphin, *Cephalorhyncus eutropia*
  - Heaviside's dolphin, *Cephalorhyncus heavisidii*
  - Hector's dolphin, *Cephalorhyncus hectori*
- Genus *Delphinus*: common dolphins
  - Long-beaked common dolphin, *Delphinus capensis*
  - Short-beaked common dolphin, *Delphinus delphis*
  - Arabian common dolphin, *Delphinus tropicalis*
- Genus *Feresa*
  - Pygmy killer whale, *Feresa attenuata*
- Genus *Globicephala*: pilot whales
  - Short-finned pilot whale, *Globicephala macrorhyncus*
  - Long-finned pilot whale, *Globicephala melas*
- Genus *Grampus*
  - Risso's dolphin, *Grampus griseus*
- Genus *Lagenodelphis*
  - Fraser's dolphin, *Lagenodelphis hosei*
- Genus *Lagenorhynchus*: false bottlenose dolphins
  - Atlantic white-sided dolphin, *Lagenorhynchus acutus*
  - White-beaked dolphin, *Lagenorhynchus albirostris*
  - Peale's dolphin, *Lagenorhynchus australis*
  - Hourglass dolphin, *Lagenorhynchus cruciger*
  - Pacific white-sided dolphin, *Lagenorhynchus obliquidens*
  - Dusky dolphin, *Lagenorhynchus obscurus*
- Genus *Lissodelphis*: right whale dolphin
  - Northern right whale dolphin, *Lissodelphis borealis*

- Southern right whale dolphin, *Lissodelphis peronii*
- Genus *Orcaella*: Irrawaddy dolphins
    - Irrawaddy dolphin, *Orcaella brevirostris*
    - Australian snubfin dolphin, *Orcaella heinsohni*
- Genus *Orcinus*
    - Killer whale, *Orcinus orca*
- Genus *Peponocephala*
    - Melon-headed whale, *Peponocephala electra*
- Genus *Pseudorca*
    - False killer whale, *Pseudorca crassidens*
- Genus *Sotalia*: northern South American dolphins
    - Tucuxi, *Sotalia fluviatilis*
    - Guiana dolphin, *Sotalia guianensis*
- Genus *Sousa*: humpback dolphins
    - Pacific humpback dolphin, *Sousa chinensis*
    - Indian humpback dolphin, *Sousa plumbea*
    - Atlantic humpback dolphin, *Sousa teuszii*
- Genus *Stenella*: spotted dolphins
    - Pantropical spotted dolphin, *Stenella attenuata*
    - Clymene dolphin, *Stenella clymene*
    - Striped dolphin, *Stenella coeruleoalba*
    - Atlantic spotted dolphin, *Stenella frontalis*
    - Spinner dolphin, *Stenella longirostris*
- Genus *Steno*
    - Rough-toothed dolphin, *Steno bredanensis*
- Genus *Tursiops*: true bottlenose dolphins
    - Indian Ocean bottlenose dolphin, *Tursiops aduncus*
    - Burrunan dolphin, *Tursiops australis*
    - Common bottlenose dolphin, *Tursiops truncatus*
- Family Monodontidae: Arctic whales

- Genus *Delphinapterus*
  - Beluga, *Delphinapterus leucas*
- Genus *Monodon*
  - Narwhal, *Monodon monoceros*
- Family Phocoenidae: Porpoises
  - Genus *Neophocaena*
    - Finless porpoise, *Neophocaena phocaenoides*
  - Genus *Phocoena*
    - Spectacled porpoise, *Phocoena dioptrica*
    - Harbour porpoise, *Phocoena phocaena*
    - Vaquita, *Phocoena sinus*
    - Burmeister's porpoise, *Phocoena spinipinnis*
  - Genus *Phocoenoides*
    - Dall's porpoise, *Phocoenoides dalli*
- Superfamily Physeteroidea: Sperm whales
  - Family Physeteridae
    - Genus *Physeter*
      - Sperm whale, *Physeter catodon* (syn. *P. macrocephalus*)
  - Family Kogiidae: lesser sperm whales
    - Genus *Kogia*
      - Pygmy sperm whale, *Kogia breviceps*
      - Dwarf sperm whale, *Kogia sima*
- Superfamily Platanistoidea: Indian river dolphins
  - Family Platanistidae
    - Genus *Platanista*
      - South Asian river dolphin, *Platanista gangetica*
- Superfamily Inioidea: South American river dolphins
  - Family Iniidae
    - Genus *Inia*

- Amazon river dolphin, *Inia geoffrensis*
- Bolivian river dolphin, *Inia boliviensis*
- Araguaian river dolphin. *Inia araguaiaensis*

- Family Pontoporiidae
  - Genus *Pontoporia*
    - La Plata dolphin, *Pontoporia blainvillei*

- Superfamily Ziphioidea: Beaked whales
  - Family Ziphidae
    - Genus *Berardius*: giant beaked whales
      - Arnoux's beaked whale, *Berardius arnuxii*
      - Baird's beaked whale (North Pacific bottlenose whale), *Berardius bairdii*
    - Subfamily Hyperoodontidae
      - Genus *Hyperoodon*: bottlenose whales
        - Northern bottlenose whale, *Hyperoodon ampullatus*
        - Southern bottlenose whale, *Hyperoodon planifrons*
      - Genus *Indopacetus*
        - Indo-Pacific beaked whale (Longman's beaked whale), *Indopacetus pacificus*
      - Genus *Mesoplodon*, Mesoplodont whale
        - Sowerby's beaked whale, *Mesoplodon bidens*
        - Andrews' beaked whale, *Mesoplodon bowdoini*
        - Hubbs' beaked whale, *Mesoplodon carlhubbsi*
        - Blainville's beaked whale, *Mesoplodon densirostris*
        - Gervais' beaked whale, *Mesoplodon europaeus*
        - Ginkgo-toothed beaked whale, *Mesoplodon ginkgodens*

- Gray's beaked whale, *Mesoplodon grayi*
- Hector's beaked whale, *Mesoplodon hectori*
- Strap-toothed whale, *Mesoplodon layardii*
- True's beaked whale, *Mesoplodon mirus*
- Perrin's beaked whale, *Mesoplodon perrini*
- Pygmy beaked whale, *Mesoplodon peruvianus*
- Stejneger's beaked whale, *Mesoplodon stejnegeri*
- Spade-toothed whale, *Mesoplodon traversii*
- Deraniyagala's beaked whale, *Mesoplodon hotaula*

- Genus *Tasmacetus*
  - Shepherd's beaked whale, *Tasmacetus shepherdi*
- Genus *Ziphius*
  - Cuvier's beaked whale, *Ziphius cavirostris*

Status

Threats

The primary threats to cetaceans come from people, both directly from whaling or drive hunting and indirect threats from fishing and pollution.

Whaling

Methods of Whaling

Japanese research ship whaling mother and calf minke whales.

An Atlantic White-sided Dolphin caught in a drive hunt in Hvalba on the Faroe Islands being taken away with a forklift.

Whaling is the practice of hunting whales, mainly baleen and sperm whales. This activity has gone on since the Stone Age.

In the Middle Ages, reasons for whaling included their meat, oil usable as fuel and the jawbone, which was used in house construction. At the end of the Middle Ages, early whaling fleets aimed at baleen whales, such as bowheads. In the 16th and 17th centuries, the Dutch fleet had about 300 whaling ships with 18,000 crewmen.

In the 18th and 19th centuries, baleen whales especially were hunted for their baleen, which was used as a replacement for wood, or in products requiring strength and flexibility such as corsets and crinoline skirts. In addition, the spermaceti found in the sperm whale was used as a machine lubricant and the ambergris as a material for pharmaceutical and perfume industries. In the second half of the 19th century, the explosive harpoon was invented, leading to a massive increase in the catch size.

Large ships were used as "mother" ships for the whale handlers. In the first half of the 20th century, whales were of great importance as a supplier of raw materials. Whales were intensively hunted during this time; in the 1930s, 30,000 whales were killed. This increased to over 40,000 animals per year up to the 1960s, when stocks of large baleen whales collapsed.

Most hunted whales are now threatened, with some great whale populations exploited to the brink of extinction. Atlantic and Korean gray whale populations were completely eradicated and the North Atlantic right whale population fell to some 300-600. The blue whale population is estimated to be around 14,000.

The first efforts to protect whales came in 1931. Some particularly endangered species, such as the humpback whale (which then numbered about 100 animals), were placed under international protection and the first protected areas were established. In 1946, the International Whaling Commission (IWC) was established, to monitor and secure whale stocks. Whaling for commercial purposes was prohibited worldwide by this organization from 1985 to 2005.

The stocks of species such as humpback and blue whales have recovered, though they are still threatened. The United States Congress passed the Marine Mammal Protection Act of 1972 sustain the marine mammal population. It prohibits the taking of marine mammals. Japanese whaling ships are allowed to hunt whales of different species for ostensibly scientific purposes. Aboriginal whaling is still permitted, but under limited circumstances as defined by IWC. Iceland and Norway do not recognize the ban and operate commercial whaling. Norway and Japan are committed to ending the ban.

Dolphins and other smaller cetaceans are hunted in an activity known as dolphin drive hunting. This is accomplished by driving a pod together with boats, usually into a bay or onto a beach. Their escape is prevented by closing off the route to the ocean with other boats or nets. Dolphins are hunted this way in several places around the world, including the Solomon Islands, the Faroe Islands, Peru and Japan (the most well-known practitioner). Dolphins are mostly hunted for their meat, though some end up in dolphinaria. Despite the controversy thousands of dolphins are caught in drive hunts each year.

Fishing

Dominoes made of baleen

Dolphin pods often reside near large tuna shoals. This is known to fishermen, who look for dolphins to catch tuna. Dolphins are much easier to spot from a distance than tuna, since they regularly breathe. The fishermen pull their nets hundreds of meters wide in a circle around the dolphin groups, in the expectation that they will net a tuna shoal. When the nets are pulled together, the dolphins become entangled under water and drown. Line fisheries in larger rivers are threats to river dolphins.

A greater threat than by-catch for small cetaceans is targeted hunting. In Southeast Asia, they are sold as fish-replacement to locals, since the region's edible fish promise higher revenues from exports. In the Mediterranean, small cetaceans are targeted to ease pressure on edible fish.

Strandings

A stranding is when a cetacean leaves the water to lie on a beach. In some cases, groups of whales strand together. The best known are mass strandings of pilot whales and sperm whales. Stranded cetaceans usually die, because their as much as 90 metric tons (99 short tons) body weight compresses their lungs or breaks their ribs. Smaller whales can die of heatstroke because of their thermal insulation.

The causes are not clear. Possible reasons for mass beachings are:

- toxic contaminants

- debilitating parasites (in the respiratory tract, brain or middle ear)

- infections (bacterial or viral)

- flight from predators (including humans)

- social bonds within a group, so that the pod follows a stranded animal

- disturbance of their magnetic senses by natural anomalies in the Earth's magnetic field

- injuries

- noise pollution by shipping traffic, seismic surveys and military sonar experiments

Since 2000, whale strandings frequently occurred following military sonar testing. In December 2001, the US Navy admitted partial responsibility for the beaching and the deaths of several marine mammals in March 2000. The coauthor of the interim report stated that animals killed by active sonar of some Navy ships were injured. Generally, underwater noise, which is still on the increase, is increasingly tied to strandings; because it impairs communication and sense of direction.

Climate change influences the major wind systems and ocean currents, which also lead to cetacean strandings. Researchers studying strandings on the Tasmanian coast from 1920-2002 found that greater strandings occurred at certain time intervals. Years with increased strandings were associated with severe storms, which initiated cold water flows close to the coast. In nutrient-rich, cold water, cetaceans expect large prey animals, so they follow the cold water currents into shallower waters, where the risk is higher for strandings. Whales and dolphins who live in pods may accompany sick or debilitated pod members into shallow water, stranding them at low tide. Once stranded, large whales are crushed by their own body weight, if they cannot quickly return to the water. In addition, body temperature regulation is compromised.

Environmental Hazards

Worldwide, use of active sonar has been linked to about 50 marine mammal strandings between 1996 and 2006. In all of these occurrences, there were other contributing factors, such as unusual (steep and complex) underwater geography, limited egress routes, and a specific species of marine mammal — beaked whales — that are suspected to be more sensitive to sound than other marine mammals.

—Rear Admiral Lawrence Rice

Heavy metals, residues of many plant and insect venoms and plastic waste flotsam are not biodegradable. Sometimes, cetaceans consume these hazardous materials, mistaking them for food items. As a result, the animals are more susceptible to disease and have fewer offspring.

Damage to the ozone layer reduces plankton reproduction because of its resulting radiation. This shrinks the food supply for many marine animals, but the filter-feeding baleen whales are most impacted. Even the Nekton is, in addition to intensive exploitation, damaged by the radiation.

Food supplies are also reduced long-term by ocean acidification due to increased absorption of increased atmospheric carbon dioxide. The CO^2 reacts with water to form carbonic acid, which reduces the construction of the calcium carbonate skeletons of food supplies for zooplankton that baleen whales depend on.

The military and resource extraction industries operate strong sonar and blasting operations. Marine seismic surveys use loud, low-frequency sound that show what is lying underneath the Earth's surface. Vessel traffic also increases noise in the oceans. Such noise can disrupt cetacean behavior such as their use of biosonar for orientation and communication. Severe instances can panic them, driving them to the surface. This leads to bubbles in blood gases and can cause decompression sickness. Naval exercises with sonar regularly results in fallen cetaceans that wash up with fatal decompression. Sounds can be disruptive at distances of more than 100 kilometres (62 mi). Damage varies across frequency and species.

Relationship to Humans

Research History

A whale as depicted by Conrad Gesner, 1587, in *Historiae animalium*

In Aristotle's time, the 4th century BCE, whales were regarded as fish due to their superficial similarity. Aristotle, however, observed many physiological and anatomical similarities with the terrestrial vertebrates, such as blood (circulation), lungs, uterus and fin anatomy. His detailed descriptions were assimilated by the Romans, but mixed with a more accurate knowledge of the dolphins, as mentioned by Pliny the Elder in his *Natural history*. In the art of this and subsequent periods, dolphins are portrayed with a high-arched head (typical of porpoises) and a long snout. The harbour porpoise was one of the most accessible species for early cetologists; because it could be seen close to land, inhabiting shallow coastal areas of Europe. Much of the findings that apply to all cetaceans were first discovered in porpoises. One of the first anatomical descriptions of the airways of a harbor porpoise dates from 1671 by John Ray. It nevertheless referred to the porpoise as a fish.

The tube in the head, through which this kind fish takes its breath and spitting water, located in front of the brain and ends outwardly in a simple hole, but inside it is divided by a downward bony septum, as if it were two nostrils; but underneath it opens up again in the mouth in a void.

—*John Ray, 1671, the earliest description of cetacean airways*

In the 10th edition of Systema Naturae (1758), Swedish biologist and taxonomist Carl Linnaeus asserted that cetaceans were mammals and not fish. His groundbreaking binomial system formed the basis of modern whale classification.

Culture

Cetaceans play a role in human culture.

Prehistoric

Stone Age petroglyphs , such as those in Roddoy and Reppa (Norway), depict them. Whale bones were used for many purposes. In the Neolithic settlement of Skara Brae on Orkney sauce pans were made from whale vertebrae.

Antiquity

"Destruction of Leviathan" engraving by Gustave Doré, 1865

Silver coin with Tarus riding a dolphin

The whale was first mentioned in ancient Greece by Homer. There, it is called Ketos, a term that initially included all large marine animals. From this was derived the Roman word for whale, Cetus. Other names were phálaina (Aristotle, Latin form of ballaena) for the female and, with an ironic characteristic style, musculus (Mouse) for the male. North Sea whales were called Physter, which was meant for the sperm whale *Physter macrocephalus*. Whales are described in particular by Aristotle, Pliny and Ambrose. All mention both live birth and suckling. Pliny describes the problems associated with the lungs with spray tubes and Ambrose claimed that large whales would take their young into their mouth to protect them.

In the Bible especially, the leviathan plays a role as a sea monster. The essence, which features a giant crocodile or a dragon and a whale, was created according to the Bible by God (Psalms 104:26) and should again be destroyed by him (Psalms 74:14 and Isaiah 27:1). In the Book of Job, the leviathan is described in more detail (Job 40:25 to Job 41:26).

In Jonah 2:1-Jonah 2:11 is a more recognizable description of a whale alongside the prophet Jonah, who, on his flight from the city of Nineveh is swallowed by a whale.

Dolphins are mentioned far more often than whales. Aristotle discusses the sacred animals of the Greeks in his *Historia Animalium* and gives details of their role as aquatic animals. The Greeks admired the dolphin as a "king of the aquatic animals" and referred to them erroneously as fish. Its intelligence was apparent both in its ability to escape from fishnets and in its collaboration with fishermen.

River dolphins are known from the Ganges and - erroneously - the Nile. In the latter case it was equated with sharks and catfish. Supposedly they attacked even crocodiles.

Dolphins appear in Greek mythology. Because of their intelligence, they rescued multiple people from drowning. They were said to love music - probably not least because of their own song - they saved, in the legends, famous musicians such as Arion of Lesbos from Methymna or Kairanos from Miletus. Because of their mental faculties, dolphins were considered for the god Dionysus.

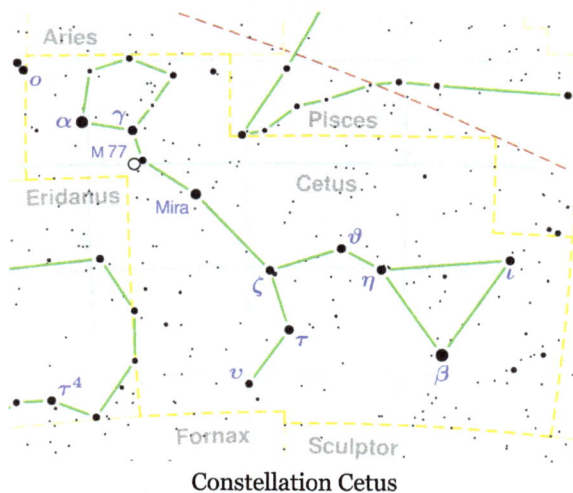

Constellation Cetus

Dolphins belong to the domain of Poseidon and led him to his wife Amphitrite. Dolphins are associated with other gods, such as Apollo, Dionysus and Aphrodite. The Greeks paid tribute to both whales and dolphins with their own constellation. The constellation of the Whale (Ketos, lat. Cetus) is located south of the Dolphin (Delphi, lat. Delphinus) north of the zodiac.

Ancient art often included dolphin representations, including the Cretan Minoans. Later they appeared on reliefs, gems, lamps, coins, mosaics and gravestones. A particularly popular representation is that of Arion or the Taras (mythology) riding on a dolphin. In early Christian art, the dolphin is a popular motif, at times used as a symbol of Christ.

Middle Ages to the 19th Century

St. Brendan described in his travel story *Navigatio Sancti Brendani* an encounter with a whale, between the years 565–573. He described how he and his companions entered a treeless island, which turned out to be a giant whale, which he called Jasconicus. He met this whale seven years later and rested on his back.

Most descriptions of large whales from this time until the whaling era, beginning in the 17th century, were of beached whales, which resembled no other animal. This was particularly true for the sperm whale, the most frequently stranded in larger groups. Raymond Gilmore documented seventeen sperm whales in the estuary of the Elbe from 1723 to 1959 and thirty-one animals on the coast of Great Britain in 1784. In 1827, a blue whale beached itself off the coast of Ostend. Whales were used as attractions in museums and traveling exhibitions.

Depiction of baleen whaling, 1840

Stranded sperm whale engraving, 1598

Whalers in the 17–19th centuries depicted whales in drawings and recounted tales of their occupation. Although they knew that whales were harmless giants, they described battles with harpooned animals. These included descriptions of sea monsters, including huge whales, sharks, sea snakes, giant squid and octopuses.

Among the first whalers who described their experiences on whaling trips was Captain William Scoresby from Great Britain, who published the book *Northern Whale Fishery*, describing the hunt for northern baleen whales. This was followed by Thomas Beale, a British surgeon, in his book *Some observations on the natural history of the sperm whale* in 1835; and Frederick Debell Bennett's *The tale of a whale hunt* in 1840. Whales were described in narrative literature and paintings, most famously in the novels *Moby Dick* by Herman Melville and *20,000 Leagues Under the Sea* by Jules Verne. In the 1882 children's book *Adventures of Pinocchio* by Carlo Collodi, the wooden figures Pinocchio and Geppettos' creators were swallowed by a whale.

Baleen was used to make vessel components such as the bottom of a bucket in the Scottish National Museum. The Norse crafted ornamented plates from baleen, sometimes interpreted as ironing boards.

In the Canadian Arctic (east coast) in Punuk and Thule culture (1000-1600 C.E.), I baleen was used to construct houses in place of wood as roof support for winter houses, with half of the building buried under the ground. The actual roof was probably made of animal skins that were covered with soil and moss.

Modern Culture

Sea World show featuring bottlenose dolphins and pilot whales

In the 20th century perceptions of cetaceans changed. They transformed from monsters into objects of wonder. As science revealed them to be intelligent and peaceful animals. Hunting was replaced by whale and dolphin tourism. This change is reflected in films and novels. For example, the protagonist of the series Flipper was a bottle-nose dolphin. The TV series SeaQuest DSV (1993-1996), the movies Free Willy, Star Trek IV: The Voyage Home and the book series The Hitchhiker's Guide to the Galaxy by Douglas Adams are examples.

The study of whale song also produced a popular Judy Collins album, Songs of the Humpback Whale.

Captivity

Whales and dolphins have been kept in captivity for use in education, research and entertainment since the 19th century.

Belugas

Beluga whales were the first whales to be kept in captivity. Other species were too rare, too shy or too big. The first was shown at Barnum's Museum in New York City in 1861. For most of the 20th century, Canada was the predominant source. They were taken from the St. Lawrence River estuary until the late 1960s, after which they were predominantly taken from the Churchill River estuary until capture was banned in 1992. Russia then became the largest provider. Belugas are caught in the Amur Darya delta and their eastern coast and are transported domestically to aquaria or dolphinaria in Moscow, St. Petersburg and Sochi, or exported to countries such as Canada. They have not been domesticated.

As of 2006, 30 belugas lived in Canada and 28 in the United States. 42 deaths in captivity had been reported. A single specimen can reportedly fetch up to US$100,000 (UK£64,160). The beluga's popularity is due to its unique color and its facial expressions. The latter is possible because while most cetacean "smiles" are fixed, the extra movement afforded by the beluga's unfused cervical vertebrae allows a greater range of apparent expression.

Killer Whales

The killer whale's intelligence, trainability, striking appearance, playfulness in captivity and sheer size have made it a popular exhibit at aquaria and aquatic theme parks. From 1976 to 1997, fifty-five whales were taken from the wild in Iceland, nineteen from Japan and three from Argentina. These figures exclude animals that died during capture. Live captures fell dramatically in the 1990s and by 1999, about 40% of the forty-eight animals on display in the world were captive-born.

Organizations such as World Animal Protection and the Whale and Dolphin Conservation Society campaign against the practice of keeping them in captivity.

In captivity, they often develop pathologies, such as the dorsal fin collapse seen in 60–90% of captive males. Captives have reduced life expectancy, on average only living into their 20s, although some live longer, including several over 30 years old and two, Corky II and Lolita, in their mid-40s. In the wild, females who survive infancy live 46 years on average and up to 70–80 years. Wild males who survive infancy live 31 years on average and can reach 50–60 years.

Captivity usually bears little resemblance to wild habitat and captive whales' social groups are foreign to those found in the wild. Critics claim captive life is stressful due to these factors and the requirement to perform circus tricks that are not part of wild killer whale behavior. Wild killer whales may travel up to 160 kilometres (100 mi) in a day and critics say the animals are too big and intelligent to be suitable for captivity. Captives occasionally act aggressively towards themselves, their tankmates, or humans, which critics say is a result of stress. Killer whales are well known for their performances in shows, but the number of orcas kept in captivity is small, especially when compared to the number of bottlenose dolphins, with only forty-four captive orcas being held in aquaria as of 2012.

Dawn Brancheau doing a show four years before the incident

Each country has its own tank requirements; in the US, the minimum enclosure size is set by the Code of Federal Regulations, 9 CFR E § 3.104, under the *Specifications for the Humane Handling, Care, Treatment and Transportation of Marine Mammals.*

Aggression among captive killer whales is common. They attack each other and their trainers as well. In 2013, SeaWorld's treatment of killer whales in captivity was the basis of the movie *Blackfish*, which documents the history of Tilikum, a killer whale at SeaWorld Orlando, who had been involved in the deaths of three people. The film was a sensation, leading the company to announce in 2016 that it would phase out its killer whale program after various unsuccessful attempts to restore its reputation and stock price.

Others

SeaWorld pilot whale with trainers

Dolphins and porpoises are kept in captivity. Bottlenose dolphins are the most common, as they are relatively easy to train, have a long lifespan in captivity and have a friendly appearance. Bottlenose dolphins live in captivity across the world, though exact numbers are hard to determine. Other species kept in captivity are spotted dolphins, false killer whales and common dolphins, Commerson's dolphins, as well as rough-toothed dolphins, but all in much lower numbers. There are also fewer than ten pilot whales, Amazon river dolphins, Risso's dolphins, spinner dolphins, or tucuxi in captivity. Two unusual and rare hybrid dolphins, known as wolphins, are kept at Sea Life Park in Hawaii, which is a cross between a bottlenose dolphin and a false killer whale. Also, two xommon/bottlenose hybrids reside in captivity at Discovery Cove and SeaWorld San Diego.

In repeated attempts in the 1960s and 1970s, narwhals kept in captivity died within months. A breeding pair of pygmy right whales were retained in a netted area. They were eventually released in South Africa. In 1971, SeaWorld captured a California gray whale calf in Mexico at Scammon's Lagoon. The calf, later named Gigi, was separated from her mother using a form of lasso attached to her flukes. Gigi was displayed at SeaWorld San Diego for a year. She was then released with a radio beacon affixed to her back; however, contact was lost after three weeks. Gigi was the first captive baleen whale. JJ, another gray whale calf, was kept at SeaWorld San Diego. JJ was an orphaned calf that beached itself in April 1997 and was transported two miles to SeaWorld. The 680 kilograms (1,500 lb) calf was a popular attraction and behaved normally, despite separation from his mother. A year later, the then 8,164.7 kilograms (18,000 lb) whale though smaller than average, was too big to keep in captivity, and was released on April 1, 1998. A captive Amazon river dolphin housed at Acuario de Valencia is the only trained river dolphin in captivity.

References

- Walters, Martin; Johnson, Jinny (2003). Encyclopedia of Animals. Marks and Spencer p.l.c. p. 229. ISBN 1-84273-964-6.

- c d Best, Robin (1984). Macdonald, D., ed. The Encyclopedia of Mammals. New York: Facts on File. pp. 292–298. ISBN 0-87196-871-1.

- Gournelos, Ted (2009). Popular Culture and the Future of Politics: Cultural Studies and the Tao of South Park. Rowman & Littlefield. pp. 142–43. ISBN 9780739137215

- Elias, J. S. (2007). Science Terms Made Easy: A Lexicon of Scientific Words and Their Root Language Origins. Greenwood Publishing Group. p. 157. ISBN 978-0-313-33896-0.

- Scheffer, Victor B. (1958). Seals, Sea Lions, and Walruses: A Review of the Pinnipedia. Stanford University Press. pp. 52–. ISBN 978-0-8047-0544-8.

- Karleskin, G.; Turner, R. L.; Small, J. W. (2009). Introduction to Marine Biology. Cengage Learning. p. 328. ISBN 978-0-495-56197-2.

- Renouf, D. (1991). "Sensory reception and processing in Phocidae and Otariidae". In Renouf, D. Behaviour of Pinnipeds. Chapman and Hall. p. 373. ISBN 978-0-412-30540-5.

- Costa, D. P. (2007). "Diving physiology of marine vertebrates" (PDF). Encyclopedia of Life Sciences. doi:10.1002/9780470015902.a0004230. ISBN 0-470-01617-5.

- Lavinge, D. M.; Kovacs, K. M.; Bonner, W. N. (2001). "Seals and Sea lions". In MacDonald, D. The Encyclopedia of Mammals (2nd ed.). Oxford University Press. pp. 147–55. ISBN 978-0-7607-1969-5.

- Cappozzo, H. L. (2001). "New perspectives on the behavioural ecology of pinnipeds". In Evans, P. G.; Raga, J. A. Marine Mammals: Biology and Conservation. Kluwer Academic/Plenum Publishers. p. 243. ISBN 978-0-306-46573-4.

- Nowak, R. M. (2003). Walker's Marine Mammals of the World. Johns Hopkins University Press. pp. 80–83. ISBN 978-0-8018-7343-0.

- Rosen, B. (2009). The Mythical Creatures Bible: The Definitive Guide to Legendary Beings. Sterling Publishing Company. p. 131. ISBN 978-1-4027-6536-0.

Whale and Marine Otter

A whale is a diverse group of aquatic mammals; whales encompass eight extant families. Baleen whales and marine otters have also been explained in the section. The chapter on whales and marine otters offers an insightful focus, keeping in mind the subject matter.

Whale

Whale is the common name for a widely distributed and diverse group of fully aquatic placental marine mammals. They are an informal grouping within the infraorder Cetacea, usually excluding dolphins and porpoises. Whales, dolphins and porpoises belong to the order Cetartiodactyla with even-toed ungulates and their closest living relatives are the hippopotamuses, having diverged about 40 million years ago. The two parvorders of whales, baleen whales (Mysticeti) and toothed whales (Odontoceti), are thought to have split apart around 34 million years ago. The whales comprise eight extant families: Balaenopteridae (the rorquals), Balaenidae (right whales), Cetotheriidae (the pygmy right whale), Eschrichtiidae (the gray whale), Monodontidae (belugas and narwhals), Physeteridae (the sperm whale), Kogiidae (the dwarf and pygmy sperm whale), and Ziphiidae (the beaked whales).

Whales are creatures of the open ocean; they feed, mate, give birth, suckle and raise their young at sea. So extreme is their adaptation to life underwater that they are unable to survive on land. Whales range in size from the 2.6 metres (8.5 ft) and 135 kilograms (298 lb) dwarf sperm whale to the 29.9 metres (98 ft) and 190 metric tons (210 short tons) blue whale, which is the largest creature that has ever lived. The sperm whale is the largest toothed predator on earth. Several species exhibit sexual dimorphism, in that the females are larger than males. Baleen whales have no teeth; instead they have plates of baleen, a fringe-like structure used to expel water while retaining the krill and plankton which they feed on. They use their throat pleats to expand the mouth to take in huge gulps of water. Balaenids have heads that can make up 40% of their body mass to take in water. Toothed whales, on the other hand, have conical teeth designed for catching fish or squid. Baleen whales have a well developed sense of "smell", whereas toothed whales have well-developed hearing – their hearing, that is adapted for both air and water, is so well developed that some can survive even if they are blind. Some species, such as sperm whales, are well adapted for diving to great depths to catch squid and other favoured prey.

Whales have evolved from land-living mammals. As such they must breathe air regularly, though they can remain submerged for long periods. They have blowholes (modified nostrils) located on top of their heads, through which air is taken in and expelled in the form of vapour. They are warm-blooded, and have a layer of fat, or blubber, under the skin. With streamlined fusiform bodies and two limbs that are modified into flippers, whales can travel at up to 20 knots, though they are not as flexible or agile as seals. Whales produce a great variety of vocalizations, notably the extended songs of the humpback whale. Although whales are widespread, most species prefer the colder waters of the Northern and Southern Hemispheres, and migrate to the equator to give birth. Species such as humpbacks and blue whales are capable of travelling thousands of miles without feeding. Males typically mate with multiple females every year, but females only mate every two to three years. Calves are typically born in the spring and summer months and females bear all the responsibility for raising them. Mothers of some species fast and nurse their young for a relatively long time.

Once relentlessly hunted for their products, whales are now protected by international law. The North Atlantic right whales nearly became extinct in the twentieth century, with a population low of 450, and the North Pacific gray whale population is ranked Critically Endangered by the IUCN. Besides whaling, they also face threats from bycatch and marine pollution. The meat, blubber and baleen of whales have traditionally been used by indigenous peoples of the Arctic. Whales have been depicted in various cultures worldwide, notably by the Inuit and the coastal peoples of Vietnam and Ghana, who sometimes hold whale funerals. Whales occasionally feature in literature and film, as in the great white whale of Herman Melville's *Moby Dick*. Small whales, such as belugas, are sometimes kept in captivity and trained to perform tricks, but breeding success has been poor and the animals often die within a few months of capture. Whale watching has become a form of tourism around the world.

Etymology and Definitions

The word "whale" comes from the Old English *whæl*, related to the High German *wal*. Related forms are the Old Norse *hvalr* and the Swedish/Danish *hval*. The obsolete "whalefish" has a similar derivation, indicating a time when whales were thought to be fish. Other archaic English forms include *wal, wale, whal, whalle, whaille, wheal*, etc.

The term "whale" is sometimes used interchangeably with dolphins and porpoises, acting as a synonym for Cetacea. Six species of dolphins have the word "whale" in their name, collectively known as blackfish: the killer whale, the melon-headed whale, the pygmy killer whale, the false killer whale, and the two species of pilot whales, all of which are classified under the family Delphinidae (oceanic dolphins). Each species has a different reason for it, for example, the killer whale was named "Ballena asesina" by Spanish sailors, which translates directly to "whale assassin" or "whale killer", but is more often translated to "killer whale".

Taxonomy and Evolution

The hippopotamuses are the sister clade to the Cetacea, which contains whales, dolphins and porpoises.

The whales are part of the largely terrestrial mammalian clade Laurasiatheria. Whales do not form a clade or order; the infraorder Cetacea includes dolphins and porpoises, which are not considered whales.

Cetaceans are divided into two parvorders: the largest parvorder, Mysticeti (baleen whales), is characterized by the presence of baleen, a sieve-like structure in the upper jaw made of keratin, which it uses to filter plankton, among others, from the water; Odontocetes (toothed whales) are characterized by bearing sharp teeth for hunting, as opposed to their counterparts' baleen.

Cetaceans and artiodactyls now are classified under the order Cetartiodactyla, often still referred to as Artiodactyla, which includes both whales and hippopotamuses. The hippopotamus and pygmy hippopotamus are the whale's closest terrestrial living relatives.

Mysticetes

Mysticetes are also known as baleen whales. They have a pair of blowholes side-by-side and lack teeth, which renders them incapable of catching larger prey; they instead have baleen plates which is a sieve-like structure in the upper jaw made of keratin, which it uses to filter plankton and other food from the water; this forces them to follow krill or plankton migrations. Some whales, such as the humpback, reside in the polar regions where they feed on a reliable source of schooling fish and krill. These animals rely on their well-developed flippers and tail fin to propel themselves through the water; they swim by moving their fore-flippers and tail fin up and down. Whale ribs loosely articulate with their thoracic vertebrae at the proximal end, but do not form a rigid rib cage. This adaptation allows their chest to compress during deep dives as the pressure increases with depth. Mysticetes consist of four families: rorquals (balaenopterids), cetotheriids, right whales (balaenids), and gray whales (eschrichtiids).

The main difference between each family of mysticete is in their feeding adaptations and subsequent behaviour. Balaenopterids are the rorquals. These animals, along with the cetotheriids, rely on their throat pleats to gulp large amounts of water while feeding. The throat pleats extend from the mouth to the navel and allow the mouth to expand to a large volume for more efficient capture of the small animals they feed on. Balaenopterids consist of two genera and eight species. Balaenids are the right whales. These animals have very large heads, which can make up as much as 40% of their body mass, and much of the head is the mouth. This allows them to take in large amounts of water into their mouths, letting them feed more effectively. Eschrichtiids have one living member: the gray whale. They are bottom feeders, mainly eating crustaceans and benthic invertebrates. They feed by turning on their sides and taking in water mixed with sediment, which is then expelled through the baleen, leaving their prey trapped inside. This is an efficient method of hunting, in which the whale has no major competitors.

Odontocetes

Odontocetes are known as toothed whales; they have teeth and only one blowhole. They rely on their well-developed sonar to find their way in the water. Toothed whales send out ultrasonic clicks using the melon. Sound waves travel through the water. Upon striking an object in the water, the sound waves bounce back at the whale. These vibrations are received through fatty tissues in the jaw, which is then rerouted into the ear-bone and into the brain where the vibrations are interpreted. All toothed whales are opportunistic, meaning they will eat anything they can fit in their throat because they are unable to chew. These animals rely on their well-developed flippers and tail fin to propel themselves through the water; they swim by moving their fore-flippers and tail fin up and down. Whale ribs loosely articulate with their thoracic vertebrae at the proximal end, but they do not form a rigid rib cage. This adaptation allows the chest to compress during deep dives as opposed to resisting the force of water pressure. Excluding dolphins and porpoises, odontocetes consist of four families: belugas and narwhals (monodontids), sperm whales (physeterids), dwarf and pygmy sperm whales (kogiids), and beaked whales (ziphiids). There are six species, sometimes referred to as "blackfish", that are dolphins commonly misconceived as whales: the killer whale, the melon-headed whale, the pygmy killer whale, the false killer whale, and the two species of pilot whales, all of which are classified under the family Delphinidae (oceanic dolphins).

The differences between families of odontocetes include size, feeding adaptations and distribution. Monodontids consist of two species: the beluga and the narwhal. They both reside in the frigid arctic and both have large amounts of blubber. Belugas, being white, hunt in large pods near the surface and around pack ice, their coloration acting as camouflage. Narwhals, being black, hunt in large pods in the aphotic zone, but their underbelly still remains white to remain camouflaged when something is looking directly up or down at them. They have no dorsal fin to prevent collision with pack ice. Physeterids and Kogiids consist of sperm whales. Sperm whales consist the largest and smallest odontocetes, and spend a large portion of their life hunting squid. *P. macrocephalus* spends most of its life in search of squid in the depths; these animals do not require any degree of light at all, in fact, blind sperm whales have been caught in perfect health. The behaviour of Kogiids remains largely unknown, but, due to their small lungs, they are thought to hunt in the photic zone. Ziphiids consist of 22 species of beaked whale. These vary from size, to coloration, to distribution, but they all share a similar hunting style. They use a suction technique, aided by a pair of grooves on the underside of their head, not unlike the throat pleats on the rorquals, to feed.

Evolution

Basilosaurus skeleton

Whales are descendants of land-dwelling mammals of the artiodactyl order (even-toed ungulates). They are related to the *Indohyus*, an extinct chevrotain-like ungulate, from which they split approximately 48 million years ago. Primitive cetaceans, or archaeocetes, first took to the sea approximately 49 million years ago and became fully aquatic 5–10 million years later. What defines an archaeocete is the presence of anatomical features exclusive to cetaceans, alongside other primitive features not found in modern cetaceans, such as visible legs or asymmetrical teeth. Their features became adapted for living in the marine environment. Major anatomical changes included their hearing set-up that channeled vibrations from the jaw to the earbone (*Ambulocetus* 49 mya), a streamlined body and the growth of flukes on the tail (*Protocetus* 43 mya), the migration of the nostrils toward the top of the cranium (blowholes), and the modification of the forelimbs into flippers (*Basilosaurus* 35 mya), and the shrinking and eventual disappearance of the hind limbs (the first odontocetes and mysticetes 34 mya).

Today, the closest living relatives of cetaceans are the hippopotamuses; these share a semi-aquatic ancestor that branched off from other artiodactyls some 60 mya. Around 40 mya, a common ancestor between the two branched off into cetacea and anthracotheres; nearly all anthracotheres became extinct at the end of the Pleistocene 2.5 mya, eventually leaving only one surviving lineage - the hippo.

Whales split into two separate parvorders around 34 mya - the baleen whales (Mysticetes) and the toothed whales (Odontocetes).

Biology

Anatomy

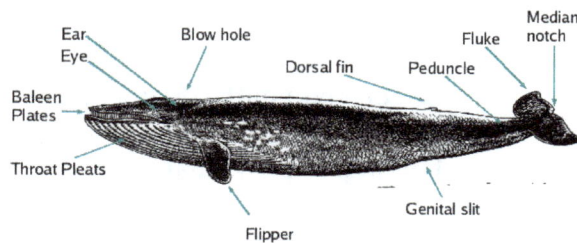

Features of a blue whale

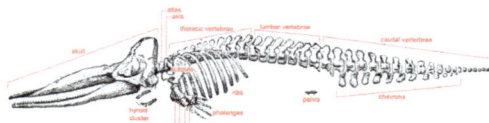

Features of a sperm whale skeleton

Whales have torpedo shaped bodies with non-flexible necks, limbs modified into flippers, non-existent external ear flaps, a large tail fin, and flat heads (with the exception of monodontids and ziphiids). Whale skulls have small eye orbits, long snouts (with the exception of monodontids and ziphiids) and eyes placed on the sides of its head. Whales range in size from the 2.6-metre (8.5 ft) and 135-kilogram (298 lb) dwarf sperm whale to the 34-metre (112 ft) and 190-metric-ton (210-short-ton) blue whale. Overall, they tend to dwarf other cetartiodactyls; the blue whale is the largest creature on earth. Several species have female-biased sexual dimorphism, with the females being larger than the males. One exception is with the sperm whale, which has males larger than the females.

Odontocetes, such as the sperm whale, possess teeth with cementum cells overlying dentine cells. Unlike human teeth, which are composed mostly of enamel on the portion of the tooth outside of the gum, whale teeth have cementum outside the gum. Only in larger whales, where the cementum is worn away on the tip of the tooth, does enamel show. Mysticetes have large whalebone, as opposed to teeth, made of keratin. Mysticetes have two blowholes, whereas Odontocetes contain only one.

Breathing involves expelling stale air from the blowhole, forming an upward, steamy spout, followed by inhaling fresh air into the lungs; a humpback whale's lungs can hold about 5,000 litres of air. Spout shapes differ among species, which facilitates identification.

All whales have a thick layer of blubber. In species that live near the poles, the blubber can be as thick as 11 inches. This blubber can help with buoyancy (which is helpful for a 100-ton whale), protection to some extent as predators would have a hard time getting through a thick layer of fat, and energy for fasting when migrating to the equator; the primary usage for blubber is insulation from the harsh climate. It can constitute as much as 50% of a whales body weight. Calves are born with only a thin layer of blubber, but some species compensate for this with thick lanugos.

Whales have a two- to three-chambered stomach that is similar in structure to terrestrial carnivores. Mysticetes contain a proventriculus as an extension of the oesophagus; this contains stones that grind up food. They also have fundic and pyloric chambers.

Locomotion

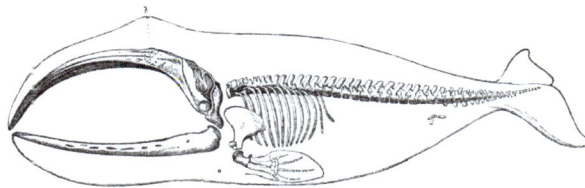

Skeleton of a bowhead whale; notice the hind limb. Richard Lydekker, 1894

Whales have two flippers on the front, and a tail fin. These flippers contain four digits. Although whales do not possess fully developed hind limbs, some, such as the sperm whale and bowhead whale, possess discrete rudimentary appendages, which may contain feet and digits. Whales are fast swimmers in comparison to seals, which typically cruise at 5–15 kn, or 9–28 kilometres per hour (5.6–17.4 mph); the fin whale, in comparison, can travel at speeds up to 47 kilometres per hour (29 mph) and the sperm whale can reach speeds of 35 kilometres per hour (22 mph). The fusing of the neck vertebrae, while increasing stability when swimming at high speeds, decreases flexibility; whales are unable to turn their heads. When swimming, whales rely on their tail fin propel them through the water. Flipper movement is continuous. Whales swim by moving their tail fin and lower body up and down, propelling themselves through vertical movement, while their flippers are mainly used for steering. Some species log out of the water, which may allow then to travel faster. Their skeletal anatomy allows them to be fast swimmers. Most species have a dorsal fin.

Whales are adapted for diving to great depths. In addition to their streamlined bodies, they can slow their heart rate to conserve oxygen; blood is rerouted from tissue tolerant of water pressure to the heart and brain among other organs; haemoglobin and myoglobin store oxygen in body tissue; and they have twice the concentration of myoglobin than haemoglobin. Before going on long dives, many whales exhibit a behaviour known as sounding; they stay close to the surface for a series of short, shallow dives while building their oxygen reserves, and then make a sounding dive.

Senses

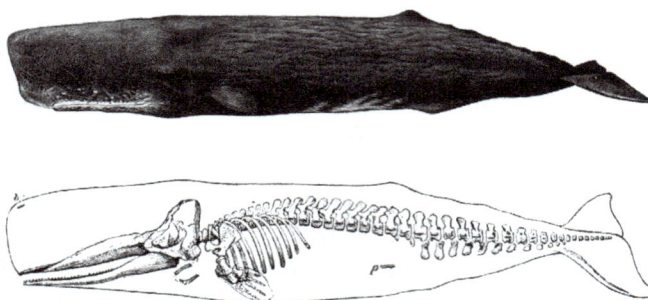

Sperm whale skeleton. Richard Lydekker, 1894.

The whale ear has specific adaptations to the marine environment. In humans, the middle ear works as an impedance equalizer between the outside air's low impedance and the cochlear fluid's high impedance. In whales, and other marine mammals, there is no great difference between the outer and inner environments. Instead of sound passing through the outer ear to the middle ear, whales receive sound through the throat, from which it passes through a low-impedance fat-filled cavity to the inner ear. The whale ear is acoustically isolated from the skull by air-filled sinus pockets, which allow for greater directional hearing underwater. Odontocetes send out high frequency clicks from an organ known as a melon. This melon consists of fat, and the skull of any such creature containing a melon will have a large depression. The melon size varies between species, the bigger the more dependent they are of it. A beaked whale for example has a small bulge sitting on top of its skull, whereas a sperm whale's head is filled up mainly with the melon.

The whale eye is relatively small for its size, yet they do retain a good degree of eyesight. As well as this, the eyes of a whale are placed on the sides of its head, so their vision consists of two fields, rather than a binocular view like humans have. When belugas surface, their lens and cornea correct the nearsightedness that results from the refraction of light; they contain both rod and cone cells, meaning they can see in both dim and bright light, but they have far more rod cells than they do cone cells. Whales do, however, lack short wavelength sensitive visual pigments in their cone cells indicating a more limited capacity for colour vision than most mammals. Most whales have slightly flattened eyeballs, enlarged pupils (which shrink as they surface to prevent damage), slightly flattened corneas and a tapetum lucidum; these adaptations allow for large amounts of light to pass through the eye and, therefore, a very clear image of the surrounding area. In water, a whale can see around 10.7 metres (35 ft) ahead of itself, but, of course, they have a smaller range above water. They also have glands on the eyelids and outer corneal layer that act as protection for the cornea.

The olfactory lobes are absent in toothed whales, suggesting that they have no sense of smell. Some whales, such as the bowhead whale, possess a vomeronasal organ, which does mean that they can "sniff out" krill.

Whales are not thought to have a good sense of taste, as their taste buds are atrophied or missing altogether. However, some toothed whales have preferences between different kinds of fish, indicating some sort of attachment to taste. The presence of the Jacobson's organ indicates that whales can smell food once inside their mouth, which might be similar to the sensation of taste.

Communication

Whale vocalization is likely to serve several purposes. Some species, such as the humpback whale, communicate using melodic sounds, known as whale song. These sounds may be extremely loud, depending on the species. Humpback whales only have been heard making clicks, while toothed whales use sonar that may generate up to 20,000 watts of sound (+73 dBm or +43 dBw) and be heard for many miles.

Captive whales have occasionally been known to mimic human speech. Scientists have suggested this indicates a strong desire on behalf of the whales to communicate with humans, as whales have a very different vocal mechanism, so imitating human speech likely takes considerable effort.

Whales emit two distinct kinds of acoustic signals, which are called whistles and clicks: Clicks are quick broadband burst pulses, used for sonar, although some lower-frequency broadband vocalizations may serve a non-echolocative purpose such as communication; for example, the pulsed calls of belugas. Pulses in a click train are emitted at intervals of ~35–50 milliseconds, and in general these inter-click intervals are slightly greater than the round-trip time of sound to the target. Whistles are narrow-band frequency modulated (FM) signals, used for communicative purposes, such as contact calls.

Intelligence

Whales are known to teach, learn, cooperate, scheme, and grieve. The neocortex of many species of whale is home to elongated spindle neurons that, prior to 2007, were known only in hominids. In humans, these cells are involved in social conduct, emotions, judgement, and theory of mind. Whale spindle neurons are found in areas of the brain that are homologous to where they are found in humans, suggesting that they perform a similar function.

Brain size was previously considered a major indicator of the intelligence of an animal. Since most of the brain is used for maintaining bodily functions, greater ratios of brain to body mass may increase the amount of brain mass available for more complex cognitive tasks. Allometric analysis indicates that mammalian brain size scales at approximately the ⅔ or ¾ exponent of the body mass. Comparison of a particular animal's brain size with the expected brain size based on such allometric analysis provides an encephalisation quotient that can be used as another indication of animal intelligence. Sperm whales have the largest brain mass of any animal on earth, averaging 8,000 cubic centimetres (490 in³) and 7.8 kilograms (17 lb) in mature males, in comparison to the average human brain which averages 1,450 cubic centimetres (88 in³) in mature males. The brain to body mass ratio in some odontocetes, such as belugas and narwhals, is second only to humans.

Small whales are known to engage in complex play behaviour, which includes such things as producing stable underwater toroidal air-core vortex rings or "bubble rings". There are two main methods of bubble ring production: rapid puffing of a burst of air into the water and allowing it to rise to the surface, forming a ring, or swimming repeatedly in a circle and then stopping to inject air into the helical vortex currents thus formed. They also appear to enjoy biting the vortex-rings, so that they burst into many separate bubbles and then rise quickly to the surface. Some believe this is a means of communication. Whales are also known to produce bubble-nets for the purpose of foraging.

A southern right whale sailing

Larger whales are also thought, to some degree, to engage in play. The southern right whale, for example, elevates their tail fluke above the water, remaining in the same position for a considerable amount of time. This is known as "sailing". It appears to be a form of play and is most commonly seen off the coast of Argentina and South Africa. Humpback whales, among others, are also known to display this behaviour.

Life Cycle

Whales are fully aquatic creatures, which means that birth and courtship behaviours are very different from terrestrial and semi-aquatic creatures. Since they are unable to go onto land to calve, they deliver the baby with the fetus positioned for tail-first delivery. This prevents the baby from drowning either upon or during delivery. To feed the new-born, whales, being aquatic, must squirt the milk into the mouth of the calf. Being mammals, they, of course, have mammary glands used for nursing calves; they are weaned off at about 11 months of age. This milk contains high amounts of fat which is meant to hasten the development of blubber; it contains so much fat that it has the consistency of toothpaste. Females deliver a single calf with gestation lasting about a year, dependency until one to two years, and maturity around seven to ten years, all varying between the species. This mode of reproduction produces few offspring, but increases the survival probability of each one. Females, referred to as "cows", carry the responsibility of childcare as males, referred to as "bulls", play no part in raising calves.

Most mysticetes reside at the poles. So, to prevent the unborn calf from dying of frostbite, they migrate to calving/mating grounds. They will then stay there for a matter of months until the calf has developed enough blubber to survive the bitter temperatures of the poles. Until then, the calves will feed on the mother's fatty milk. With the exception of the humpback whale, it is largely unknown when whales migrate. Most will travel from the Arctic or Antarctic into the tropics to mate, calve, and raise during the winter and spring; they will migrate back to the poles in the warmer summer months so the calf can continue growing while the mother can continue eating, as they fast in the breeding grounds. One exception to this is the southern right whale, which migrates to Patagonia and western New Zealand to calve; both are well out of the tropic zone.

Sleep

Unlike most animals, whales are conscious breathers. All mammals sleep, but whales cannot afford to become unconscious for long because they may drown. While knowledge of sleep in wild cetaceans is limited, toothed cetaceans in captivity have been recorded to sleep with one side of their brain at a time, so that they may swim, breathe consciously, and avoid both predators and social contact during their period of rest.

A 2008 study found that sperm whales sleep in vertical postures just under the surface in passive shallow 'drift-dives', generally during the day, during which whales do not respond to passing vessels unless they are in contact, leading to the suggestion that whales possibly sleep during such dives.

Ecology

Foraging and Predation

Polar bear with the remains of a beluga

All whales are carnivorous and predatory. Odontocetes, as a whole, mostly feed on fish and cephalopods, and then followed by crustaceans and bivalves. All species are generalist and opportunistic feeders. Mysticetes, as a whole, mostly feed on krill and plankton, followed by crustaceans and other invertebrates. A few are specialists. Examples include the blue whale, which eats almost exclusively krill, the minke whale, which eats mainly schooling fish, the sperm whale, which specialize on squid, and the gray whale which feed on bottom-dwelling invertebrates. The elaborate baleen "teeth" of filter-feeding species, mysticetes, allow them to remove water before they swallow their planktonic food by using the teeth as a sieve. Usually whales hunt solitarily, but they do sometimes hunt cooperatively in small groups. The former behaviour is typical when hunting non-schooling fish, slow-moving or immobile invertebrates or endothermic prey. When large amounts of prey are available, whales such as certain mysticetes hunt cooperatively in small groups. Some cetaceans may forage with other kinds of animals, such as other species of whales or certain species of pinnipeds.

Large whales, such as mysticetes, are not usually subject to predation, but smaller whales, such as monodontids or ziphiids, are. These species are preyed on by the killer whale or orca. To subdue and kill whales, orcas continuously ram them with their heads; this can sometimes kill bowhead whales, or severely injure them. Other times they corral the narwhals or belugas before striking. They are typically hunted by groups of 10 or fewer orcas, but they are seldom attacked by an individual. Calves are more commonly taken by orcas, but adults can be targeted as well.

These small whales are also targeted by terrestrial and pagophilic predators. The polar bear is well adapted for hunting Arctic whales and calves. Bears are known to use sit-and-wait tactics as well as active stalking and pursuit of prey on ice or water. Whales lessen the chance of predation by gathering in groups. This however means less room around the breathing hole as the ice slowly closes the gap. When out at sea, whales dive out of the reach of surface-hunting orcas. Polar bear attacks on belugas and narwhals are usually successful in winter, but rarely inflict any damage in summer.

Whale Pump

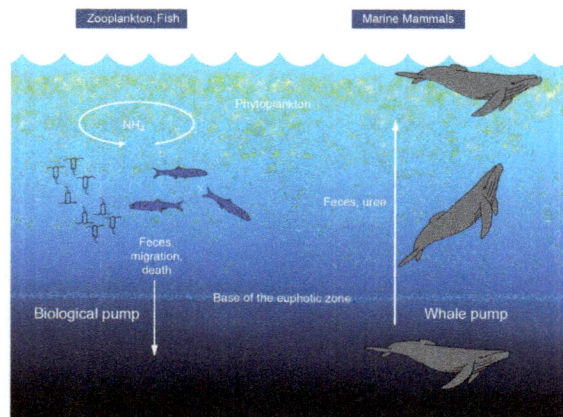

"Whale pump" – the role played by whales in recycling ocean nutrients

A 2010 study considered whales to be a positive influence to the productivity of ocean fisheries, in what has been termed a "whale pump." Whales carry nutrients such as nitrogen from the depths back to the surface. This functions as an upward biological pump, reversing an earlier presumption that whales accelerate the loss of nutrients to the bottom. This nitrogen input in the Gulf of Maine is "more than the input of all rivers combined" emptying into the gulf, some 23,000 metric tons (25,000 short tons) each year. Whales defecate at the ocean's surface; their excrement is important for fisheries because it is rich in iron and nitrogen. The whale faeces are liquid and instead of sinking, they stay at the surface where phytoplankton feed off it.

Whale Fall

Upon death, whale carcasses fall to the deep ocean and provide a substantial habitat for marine life. Evidence of whale falls in present-day and fossil records shows that deep sea whale falls support a rich assemblage of creatures, with a global diversity of 407 species, comparable to other neritic biodiversity hotspots, such as cold seeps and hydrothermal vents.

Deterioration of whale carcasses happens though a series of three stages. Initially, moving organisms such as sharks and hagfish, scavenge the soft tissues at a rapid rate over a period of months, and as long as two years. This is followed by the colonization of bones and surrounding sediments (which contain organic matter) by enrichment opportunists, such as crustaceans and polychaetes, throughout a period of years. Finally, sulfophilic bacteria reduce the bones releasing hydrogen sulphide enabling the growth of chemoautotrophic organisms, which in turn, support other organisms such as mussels, clams, limpets, and sea snails. This stage may last for decades and supports a rich assemblage of species, averaging 185 species per site.

Interaction with Humans

Whaling

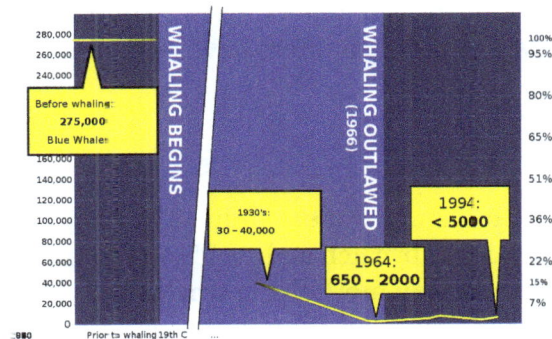

World population graph of blue whales

Whaling by humans has existed since the Stone Age. Ancient whalers used harpoons to spear the bigger animals from boats out at sea. People from Norway started hunting whales around 2000 B.C., and people from Japan began hunting whales in the Pacific at least as early. Whales are typically hunted for their meat and blubber by aboriginal groups; they used baleen for baskets or roofing, and made tools and masks out of bones. The Inuit hunted whales in the Arctic Ocean. The Basques started whaling as early as the 11th century, sailing as far as Newfoundland in the 16th century in search of right whales. 18th and 19th century whalers hunted down whales mainly for their oil, which was used as lamp fuel and a lubricant, baleen or whalebone, which was used for items such as corsets and skirt hoops, and ambergris, which was used as a fixative for perfumes. The most successful whaling nations at this time were the Netherlands, Japan, and the United States.

Commercial whaling was historically important as an industry well throughout the 17th, 18th and 19th centuries. Whaling was at that time a sizeable European industry with ships from Britain, France, Spain, Denmark, the Netherlands and Germany, sometimes collaborating to hunt whales in the Arctic, sometimes in competition leading even to war. By the early 1790s, whalers, namely the Americans and Australians, mainly focused efforts in the South Pacific where they mainly hunted sperm whales and right whales, with catches of up to 39,000 right whales by Americans alone. By 1853, U.S. profits turned to US$11,000,000 (UK£6.5m), equivalent to US$348,000,000

(UK£230m) today, the most profitable year for the American whaling industry. Commonly exploited species included North Atlantic right whales, sperm whales, which were mainly hunted by Americans, bowhead whales, which were mainly hunted by the Dutch, common minke whales, blue whales, and gray whales. The scale of whale harvesting decreased substantially after 1982 when the International Whaling Commission (IWC) placed a moratorium which set a catch limit for each country, excluding aboriginal groups until 2004.

Current whaling nations are Norway, Iceland, and Japan, despite their joining to the IWC, as well as the aboriginal communities of Siberia, Alaska, and northern Canada. Subsistence hunters typically use whale products for themselves and depend on them for survival. National and international authorities have given special treatment to aboriginal hunters since their methods of hunting are seen as less destructive and wasteful. This distinction is being questioned as these aboriginal groups are using more modern weaponry and mechanized transport to hunt with, and are selling whale products in the marketplace. Some anthropologists argue that the term "subsistence" should also apply to these cash-based exchanges as long as they take place within local production and consumption. In 1946, the IWC placed a moratorium, limiting the annual whale catch. Since then, yearly profits for these "subsistence" hunters have been close to US$31 million (UK£20m) per year.

Other Threats

Whales can also be threatened by humans more indirectly. They are unintentionally caught in fishing nets by commercial fisheries as bycatch and accidentally swallow fishing hooks. Gillnetting and Seine netting is a significant cause of mortality in whales and other marine mammals. Species commonly entangled include beaked whales. Whales are also affected by marine pollution. High levels of organic chemicals accumulate in these animals since they are high in the food chain. They have large reserves of blubber, more so for toothed whales as they are higher up the food chain than baleen whales. Lactating mothers can pass the toxins on to their young. These pollutants can cause gastrointestinal cancers and greater vulnerability to infectious diseases. They can also be poisoned by swallowing litter, such as plastic bags. Environmentalists speculate that advanced naval sonar endangers some whales. Some scientists suggest that sonar may trigger whale beachings, and they point to signs that such whales have experienced decompression sickness.

Conservation

World map showing International Whaling Commission (IWC) members in blue

The scale of whale harvesting decreased substantially after 1946 when, in response to the steep decline in whale populations, the International Whaling Commission placed a moratorium which set a catch limit for each country; this excluded aboriginal groups up until 2004. As of 2015, aboriginal communities are allowed to take 280 bowhead whales off of Alaska and two from the western coast of Greenland, 620 gray whales off of Washington state, three common minke whales off of the eastern coast of Greenland and 178 on their western coast, 10 fin whales from the west coast of Greenland, nine humpback whales from the west coast of Greenland and 20 off of St. Vincent and the Grenadines each year. Several species that were commercially exploited have rebounded in numbers; for example, Grey whales may be as numerous as they were prior to harvesting, but the North Atlantic population is functionally extinct. Conversely, the North Atlantic right whale was extirpated from much of its former range, which stretched across the North Atlantic, and only remains in small fragments along the coast of Canada, Greenland, and is considered functionally extinct along the European coastline.

The IWC has designated two whale sanctuaries: the Southern Ocean Whale Sanctuary, and the Indian Ocean Whale Sanctuary. The Southern Ocean whale sanctuary spans 30,560,860 square kilometres (11,799,610 sq mi) and envelopes Antarctica. The Indian Ocean whale sanctuary takes up all of the Indian Ocean south of 55°S. The IWC is a voluntary organization, with no treaty. Any nation may leave as they wish; the IWC cannot enforce any law it makes.

As of 2013, the International Union for Conservation of Nature (IUCN) recognized 86 cetacean species, 40 of which are considered whales. Six are considered at risk, as they are ranked Critically Endangered (the North Atlantic right whale), "Endangered" (blue whale, fin whale, North Pacific right whale, and sei whale), and "Vulnerable" (sperm whale). Twenty-one species have a "Data Deficient" ranking. Species that live in polar habitats are vulnerable to the effects of recent and ongoing climate change, particularly the time when pack ice forms and melts.

Whale Watching

Whale watching off Bar Harbour, Maine

An estimated 13 million people went whale watching globally in 2008, in all oceans except the Arctic. Rules and codes of conduct have been created to minimize harassment of the whales. Iceland, Japan and Norway have both whaling and whale watching industries. Whale watching lobbyists are concerned that the most inquisitive whales, which approach boats closely and provide much of the entertainment on whale-watching trips, will be the first to be taken if whaling is resumed in the same areas. Whale watching generated US$2.1 billion (UK£1.4 billion) per annum in tourism revenue worldwide, employing around 13,000 workers. In contrast, the whaling industry, with the moratorium in place, generates US$31 million (UK£20 million) per year. The size and rapid growth of the industry has led to complex and continuing debates with the whaling industry about the best use of whales as a natural resource.

In Myth, Literature and Art

Whalers off Twofold Bay, New South Wales. Watercolour by Oswald Brierly, 1867

As marine creatures that reside in either the depths or the poles, humans knew very little about whales over the course of history; many feared or revered them. The Nords and various arctic tribes revered the whale as they were important pieces of their lives. In Inuit creation myths, when 'Big Raven', a deity in human form, found a stranded whale, he was told by the Great Spirit where to find special mushrooms that would give him the strength to drag the whale back to the sea and thus, return order to the world. In an Icelandic legend, a man threw a stone at a fin whale and hit the blowhole, causing the whale to burst. The man was told not to go to sea for twenty years, but during the nineteenth year he went fishing and a whale came and killed him.

Whales played a major part in shaping the art forms of many coastal civilizations, such as the Norse, with some dating to the Stone Age. Petroglyphs off a cliff face in Bangudae, South Korea show 300 depictions of various animals, a third of which are whales. Some show particular detail in which there are throat pleats, typical of rorquals. These petroglyphs show these people, of around 7,000 to 3,500 B.C.E. in South Korea, had a very high dependency on whales.

The Pacific Islanders and Australian Aborigines viewed whales as bringers of good and joy. One exception is French Polynesia, where in many parts, cetaceans are met with great brutality.

In Vietnam and Ghana, among other places, whales hold a sense of divinity. They are so respected in their cultures that they occasionally hold funerals for beached whales, a throwback to Vietnam's ancient sea-based Austro-Asiatic culture. The god of the seas, according to Chinese folklore, was a large whale with human limbs.

Whales have also played a role in sacred texts such as the Bible. It mentions whales in Genesis 1:21, Job 7:12, and Ezekiel 32:2. The "leviathan" described at length in Job 41:1-34 is generally understood to refer to a whale. The "sea monsters" in Lamentations 4:3 have been taken by some to refer to marine mammals, in particular whales, although most modern versions use the word "jackals" instead. The story of Jonah being swallowed by a great fish is told both in the Qur'an and in the Bible. A medieval column capital sculpture depicting this was made in the 12th century in the abbey church in Mozac, France. The Old Testament contains the Book of Jonah and in the New Testament, Jesus mentions this story in Matthew 12:40.

In 1585, Alessandro Farnese, 1585, and Francois, Duke of Anjou, 1582, were greeted on his cere-monial entry into the port city of Antwerp by floats including "Neptune and the Whale", indicating at least the city's dependence on the sea for its wealth.

Whales continue to be prevalent in modern literature. For example, Herman Melville's *Moby Dick* features a "great white whale" as the main antagonist for Ahab, who eventually is killed by it. The whale is an albino sperm whale, considered by Melville to be the largest type of whale, and is partly based on the historically attested bull whale Mocha Dick. Rudyard Kipling's *Just So Stories* in-cludes the story of "How the Whale got in his Throat". Niki Caro's film the *Whale Rider* has a Māori girl ride a whale in her journey to be a suitable heir to the chieftain-ship. Walt Disney's film *Pinoc-chio* features a giant whale named Monstro as the final antagonist. Alan Hovhaness' orchestra *And God Created Great Whales* including the recorded sounds of humpback and bowhead whales. Léo Ferré's song "Il n'y a plus rien" is an example of biomusic that begins and ends with recorded whale songs mixed with a symphonic orchestra and his voice.

In Captivity

Beluga whales and trainers in an aquarium

Belugas were the first whales to be kept in captivity. Other species were too rare, too shy, or too big. The first beluga was shown at Barnum's Museum in New York City in 1861. For most of the 20th century, Canada was the predominant source of wild belugas. They were taken from the St. Lawrence River estuary until the late 1960s, after which they were predominantly taken from the Churchill River estuary until capture was banned in 1992. Russia has become the largest provider since it had been banned in Canada. Belugas are caught in the Amur River delta and their eastern coast, and then are either transported domestically to aquariums or dolphinariums in Moscow, St. Petersburg, and Sochi, or exported to other countries, such as Canada. Most captive belugas are caught in the wild, since captive-breeding programs are not very successful.

As of 2006, 30 belugas were in Canada and 28 in the United States, and 42 deaths in captivity had been reported up to that time. A single specimen can reportedly fetch up to US$100,000 (UK£64,160) on the market. The beluga's popularity is due to its unique colour and its facial expressions. The latter is possible because while most cetacean "smiles" are fixed, the extra movement afforded by the beluga's unfused cervical vertebrae allows a greater range of apparent expression.

Between 1960 and 1992, the Navy carried out a program that included the study of marine mammals' abilities with sonar, with the objective of improving the detection of underwater objects. A large number of belugas were used from 1975 on, the first being dolphins. The program also included training them to carry equipment and material to divers working underwater by holding cameras in their mouths to locate lost objects, survey ships and submarines, and underwater monitoring. A similar program was used by the Russian Navy during the Cold War, in which belugas were also trained for antimining operations in the Arctic.

Aquariums have tried housing other species of whales in captivity. The success of belugas turned attention to maintaining their relative, the narwhal, in captivity. However, in repeated attempts in the 1960s and 1970s, all narwhals kept in captivity died within months. A pair of pygmy right whales were retained in an enclosed area (with nets); they were eventually released in South Africa. There was one attempt to keep a stranded Sowerby's beaked whale calf in captivity; the calf rammed into the tank wall, breaking its rostrum, which resulted in death. It was thought that Sowerby's beaked whale evolved to swim fast in a straight line, and a 30 metres (98 ft) was not big enough. There have been attempts to keep baleen whales in captivity. There were three attempts to keep gray whales in captivity. Gigi was a gray whale calf that died in transport. Gigi II was another gray whale calf that was captured in the Ojo de Liebre Lagoon, and was transported to SeaWorld. The 680 kilograms (1,500 lb) calf was a popular attraction, and behaved normally, despite being separated from his mother. A year later, the 8,000-kilogram (18,000 lb) whale grew too big to keep in captivity and was released; it was the first of two gray whales, the other being another gray whale calf named JJ, to successfully be kept in captivity. There were three attempts to keep minke whales in captivity in Japan. They were kept in a tidal pool with a sea-gate at the Izu Mito Sea Paradise. Another, unsuccessful, attempt was made by the U.S. One stranded humpback whale calf was kept in captivity for rehabilitation, but died days later.

Toothed Whale

The toothed whales (systematic name Odontoceti) form a parvorder of the infraorder Cetacea, including sperm whales, beaked whales, dolphins, and porpoises. As the name suggests, the parvorder is characterized by the presence of teeth rather than the baleen of other whales. Seventy-three species of toothed whales are described. They are thought to have split from baleen whales, parvorder Mysticeti, around 34 million years ago (mya). Whales and dolphins, the paraphyletic groups of Cetacea, as well as porpoises, belong to the clade Cetartiodactyla with even-toed ungulates; their closest living relatives are the hippopotamuses which diverged about 40 mya.

Toothed whales range in size from the 4.5 ft (1.4 m) and 120 lb (54 kg) vaquita to the 20 m (66 ft) and 55 t (61-short-ton) sperm whale. Several species of odontocetes exhibit sexual dimorphism, in that the females are larger than males. They have streamlined bodies and two limbs that are modified into flippers. Some can travel at up to 20 knots. Odontocetes have conical teeth designed for catching fish or squid. They have well-developed hearing, that is well adapted for both air and water, so much so that some can survive even if they are blind. Some species are well adapted for diving to great depths. Almost all have a layer of fat, or blubber, under the skin to keep warm in the cold water, with the exception of river dolphins.

Toothed whales consist of some of the most widespread mammals, but some, as with the vaquita, are restricted to certain areas. Odontocetes feed largely on fish and squid, but a few, like the killer whale, feed on mammals, such as pinnipeds. Males typically mate with multiple females every year, but females only mate every two to three years, making them polygynous. Calves are typically born in the spring and summer, and females bear the responsibility for raising them, but more sociable species rely on the family group to care for calves. Many species, mainly dolphins, are highly sociable, with some pods reaching over a thousand individuals.

Once hunted for their products, cetaceans are now protected by international law. Some species are attributed with high levels of intelligence. At the 2012 meeting of the American Association for the Advancement of Science, support was reiterated for a cetacean bill of rights, listing cetaceans as nonhuman persons. Besides whaling and drive hunting, they also face threats from bycatch and marine pollution. The baiji, for example, is considered functionally extinct by the IUCN, with the last sighting in 2004, due to heavy pollution to the Yangtze River. Whales occasionally feature in literature and film, as in the great white sperm whale of Herman Melville's *Moby-Dick*. Small odontocetes, mainly dolphins, are kept in captivity and trained to perform tricks. Whale watching has become a form of tourism around the world.

Taxonomy

Research History

The tube in the head, through which this kind fish takes its breath and spitting water, located in front of the brain and ends outwardly in a simple hole, but inside it is divided by a downward bony septum, as if it were two nostrils; but underneath it opens up again in the mouth in a void.

—John Ray, 1671, the earliest description of cetacean airways

In Aristotle's time, the fourth century BCE, whales were regarded as fish due to their superficial similarity. Aristotle, however, could already see many physiological and anatomical similarities with the terrestrial vertebrates, such as blood (circulation), lungs, uterus, and fin anatomy. His detailed descriptions were assimilated by the Romans, but mixed with a more accurate knowledge of the dolphins, as mentioned by Pliny the Elder in his *Natural history*. In the art of this and subsequent periods, dolphins are portrayed with a high-arched head (typical of porpoises) and a long snout. The harbor porpoise is one of the most accessible species for early cetologists, because it could be seen very close to land, inhabiting shallow coastal areas of Europe. Many of the findings that apply to all cetaceans were therefore first discovered in the porpoises. One of the first anatomical descriptions of the airways of the whales on the basis of a harbor porpoise dates from 1671 by John Ray. It nevertheless referred to the porpoise as a fish.

Evolution

Toothed whales, as well as baleen whales, are descendants of land-dwelling mammals of the artiodactyl order (even-toed ungulates). They are closely related to the hippopotamus, sharing a common ancestor that lived around 54 million years ago (mya). The primitive cetaceans, or archaeocetes, first took to the sea approximately 49 mya and became fully aquatic by 5–10 million years later.

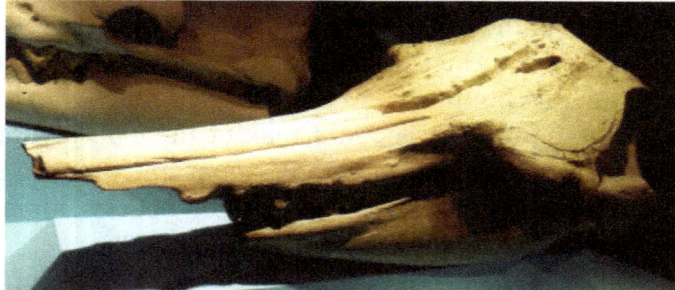

Fossil of *Squalodon*

The adaptation of echolocation occurred when toothed whales split apart from baleen whales, and distinguishes modern toothed whales from fully aquatic archaeocetes. This happened around 34 mya. Modern toothed whales do not rely on their sense of sight, but rather on their sonar to hunt prey. Echolocation also allowed toothed whales to dive deeper in search of food, with light no longer necessary for navigation, which opened up new food sources. Toothed whales (Odontocetes) echolocate by creating a series of clicks emitted at various frequencies. Sound pulses are emitted through their melon-shaped foreheads, reflected off objects, and retrieved through the lower jaw. Skulls of *Squalodon* show evidence for the first hypothesized appearance of echolocation. *Squalodon* lived from the early to middle Oligocene to the middle Miocene, around 33-14 mya. *Squalodon* featured several commonalities with modern Odontocetes. The cranium was well compressed, the rostrum telescoped outward (a characteristic of the modern parvorder Odontoceti), giving *Squalodon* an appearance similar to that of modern toothed whales. However, it is thought unlikely that squalodontids are direct ancestors of living dolphins.

Classification

- Infraorder Cetacea
  - Parvorder Odontoceti: toothed whales
    - Superfamily Delphinoidea: dolphins and relatives
    - Family Delphinidae: oceanic dolphins
    - Subfamily Delphininae
      - Genus *Delphinus*
        - Short-beaked common dolphin, *Delphinus delphis*
        - Long-beaked common dolphin, *Delphinus capensis*
        - (Arabian common dolphin, *Delphinus tropicalis*)
      - Genus *Lagenodelphis*
        - Fraser's dolphin, *Lagenodelphis hosei*

- Genus *Sousa*
  - Atlantic humpback dolphin, *Sousa teuszi*
  - Indian humpback dolphin, *Sousa plumbea*
  - Chinese white dolphin, *Sousa chinensis*
- Genus *Stenella* (syn. Clymenia, Micropia, Fretidelphis, Prodelphinus)
  - Pantropical spotted dolphin, *Stenella attenuata*
  - Atlantic spotted dolphin, *Stenella frontalis*
  - Spinner dolphin, *Stenella longirostris*
  - Clymene dolphin, *Stenella clymene*
  - Striped dolphin, *Stenella coeruleoalba*
- Genus *Tursiops*
  - Bottlenose dolphin, *Tursiops truncatus*
  - Indian Ocean bottlenose dolphin, *Tursiops aduncus*
  - Burrunan dolphin, *Tursiops australis*
- Subfamily Lissodelphininae
  - Genus *Cephalorhynchus* (syn. *Eutropia*)
  - Commerson's dolphin, *Cephalorhynchus commersonii*
  - Chilean dolphin, *Cephalorhynchus eutropia*
  - Heaviside's dolphin, *Cephalorhynchus heavisidii*
  - Hector's dolphin, *Cephalorhynchus hectori*
- Genus *Lissodelphis* (syn. *Tursio, Leucorhamphus*)
  - Northern right whale dolphin, *Lissodelphis borealis*
  - Southern right whale dolphin, *Lissodelphis peronii*
- Subfamily Orcininae
  - Genus *Feresa*
  - Pygmy killer whale, *Feresa attenuata*
- Genus *Globicephala* (syn. *Sphaerocephalus, Globiceps, Globicephalus*)
  - Long-finned pilot whale, *Globicephala melas*
  - Short-finned pilot whale, *Globicephala macrorhyncus*
  - Genus *Grampus* (syn. *Grampidelphis, Grayius*)

- Risso's dolphin, *Grampus griseus*
- Genus *Orcaella*
  - Irrawaddy dolphin, *Orcaella brevirostris*
  - Australian snubfin dolphin, *Orcaella heinsohni*
- Genus *Orcinus* (syn. *Orca, Ophysia, Gladiator*)
  - Killer whale (orca), *Orcinus orca*
- Genus *Peponocephala*
  - Melon-headed whale, *Peponocephala electra*
- Genus †*Platalearostrum* (blunt-snouted dolphin)
  - †Hoekman's blunt-snouted dolphin, *Platalearostrum hoekmani*
- Genus *Pseudorca* (syn. *Neorca*)
  - False killer whale, *Pseudorca crassidens*
- Subfamily Stenoninae
  - Genus *Sotalia* (syn. *Tucuxa*)
  - Tucuxi, *Sotalia fluviatilis*
  - Costero, *Sotalia guianensis*
  - Genus *Steno* (syn. *Glyphidelphis, Stenopontistes*)
    - Rough-toothed dolphin, *Steno bredanensis*
- Subfamily incertae sedis
  - Genus *Lagenorhynchus*
  - White-beaked dolphin, *Lagenorhynchus albirostris*
  - Atlantic white-sided dolphin, *Lagenorhynchus acutus*
  - Pacific white-sided dolphin, *Lagenorhynchus obliquidens*
  - Dusky Dolphin, *Lagenorhynchus obscurus*
  - Black-chinned dolphin, *Lagenorhynchus australis*
  - Hourglass dolphin, *Lagenorhynchus cruciger*
- Family Monodontidae
  - Subfamily Delphinapterinae
  - Genus *Delphinapterus*
  - Beluga whale, *Delphinapterus leucas*

- Subfamily Monodontinae
  - Genus *Monodon*
  - Narwhal, *Monodon monoceros*
- Family Phocoenidae: porpoises
  - Subfamily Phocoeninae
    - Genus *Neophocaena* (syn. *Meomeris*)
    - Finless porpoise, *Neophocaena phocaenoides*
    - Genus *Phocoena* (syn. *Australophocaena, Acanthodelphis*)
    - Harbour porpoise, *Phocoena phocaena*
    - Vaquita, *Phocoena sinus*
    - Spectacled porpoise, *Phocoena dioptrica*
    - Burmeister's porpoise, *Phocoena spinipinnis*
  - Subfamily Phocoenoidinae
    - Genus *Phocoenoides*
    - Dall's porpoise, *Phocoenoides dalli*
- Superfamily Inioidea, river dolphins
  - Family Iniidae
  - Genus *Inia*
  - Bolivian river dolphin, *Inia boliviensis*
  - Amazon river dolphin, *Inia geoffrensis*
  - Araguaian river dolphin, *Inia araguaiaensis*
  - Family Pontoporiidae
  - Genus *Pontoporia*
  - La Plata dolphin, *Pontoporia blainvillei*
- Superfamily Platanistoidea, river dolphins
  - Family Platanistidae
    - Genus *Platanista*
    - Ganges and Indus River dolphin, *Platanista gangetica*
  - Family †Squalodontidae
    - Genus †*Eosqualodon*

- Genus †*Macrophoca*
- Genus †*Neosqualodon*
- Genus †*Phoberodon*
- Genus †*Phocodon*
- Genus †*Smilocamptus*
- Genus †*Squalodon* (jr. synonyms *Arionius, Crenidelphinus, Kelloggia, Rhizoprion*)
- Genus †*Tangaroasaurus*
- Superfamily Lipotoidea, river dolphins (potentially extinct)
    - Family Lipotidae
    - Genus *Lipotes*
    - Chinese river dolphin, *Lipotes vexillifer*
- Superfamily Physeteroidea, sperm whales
    - Family Kogiidae
    - Genus *Kogia*
    - Dwarf sperm whale, *Kogia sima*
    - Pygmy sperm whale, *Kogia breviceps*
    - Family Physeteridae: sperm whale family
    - Genus *Physeter*
    - Sperm whale, *Physeter macrocephalus*
- Superfamily Ziphioidea, beaked whales
    - Family Ziphidae, beaked whales
    - Subfamily Berardiinae
    - Genus *Berardius*, giant beaked whales
    - Arnoux's beaked whale, *Berardius arnuxii*
    - Baird's beaked whale (North Pacific bottlenose whale), *Berardius bairdii*
    - Subfamily Hyperoodontinae
    - Genus *Hyperoodon*
        - Northern bottlenose whale, *Hyperoodon ampullatus*
        - Southern bottlenose whale, *Hyperoodon planifrons*

- Genus *Indopacetus*
  - Tropical bottlenose whale (Longman's beaked whale), *Indopacetus pacificus*
- Genus *Mesoplodon*, mesoplodont whales
  - Hector's beaked whale, *Mesoplodon hectori*
  - True's beaked whale, *Mesoplodon mirus*
  - Gervais' beaked whale, *Mesoplodon europaeus*
  - Sowerby's beaked whale, *Mesoplodon bidens*
  - Gray's beaked whale, *Mesoplodon grayi*
  - Pygmy beaked whale, *Mesoplodon peruvianus*
  - Andrews' beaked whale, *Mesoplodon bowdoini*
  - Bahamonde's beaked whale, *Mesoplodon bahamondi*
  - Hubbs' beaked whale, *Mesoplodon carlhubbsi*
  - Ginkgo-toothed beaked whale, *Mesoplodon ginkgodens*
  - Stejneger's beaked whale, *Mesoplodon stejnegeri*
  - Strap-toothed whale, *Mesoplodon layardii*
  - Blainville's beaked whale, *Mesoplodon densirostris*
  - Perrin's beaked whale, *Mesoplodon perrini*
  - Deraniyagala's beaked whale, *Mesoplodon hotaula*
- Subfamily Ziphiinae
  - Genus *Tasmacetus*
    - Shepherd's beaked whale (Shepherd's beaked whale), *Tasmacetus shepherdi*
  - Genus *Ziphius*
    - Cuvier's beaked whale, *Ziphius cavirostris*

Biology

Anatomy

Toothed whales have torpedo-shaped bodies with inflexible necks, limbs modified into flippers, nonexistent external ear flaps, a large tail fin, and bulbous heads (with the exception of sperm whales). Their skulls have small eye orbits, long beaks (with the exception sperm whales), and eyes placed on the sides of their heads. Toothed whales range in size from the 4.5 ft (1.4 m) and 120 lb (54 kg) vaquita to the 20 m (66 ft) and 55 t (61-short-ton) sperm whale. Overall, they tend to be dwarfed by their relatives, the baleen whales (Mysticeti). Several species have sexual dimorphism, with the females being larger than the males. One exception is with the sperm whale, which has males larger than the females.

Odontocetes, such as the sperm whale, possess teeth with cementum cells overlying dentine cells. Unlike human teeth, which are composed mostly of enamel on the portion of the tooth outside of the gum, whale teeth have cementum outside the gum. Only in larger whales, where the cementum is worn away on the tip of the tooth, does enamel show. Except for the sperm whale, most toothed whales are smaller than the baleen whales. The teeth differ considerably among the species. They may be numerous, with some dolphins bearing over 100 teeth in their jaws. At the other extreme are the narwhals with their single long tusks and the almost toothless beaked whales with tusk-like teeth only in males.Not all species are believed to use their teeth for feeding. For instance, the sperm whale likely uses its teeth for aggression and showmanship.

Breathing involves expelling stale air from their one blowhole, forming an upward, steamy spout, followed by inhaling fresh air into the lungs. Spout shapes differ among species, which facilitates identification. The spout only forms when warm air from the lungs meets cold air, so it does not form in warmer climates, as with river dolphins.

Almost all cetaceans have a thick layer of blubber, with the exception of river dolphins. In species that live near the poles, the blubber can be as thick as 11 in (28 cm). This blubber can help with buoyancy, protection to some extent as predators would have a hard time getting through a thick layer of fat, energy for fasting during leaner times, and insulation from the harsh climates. Calves are born with only a thin layer of blubber, but some species compensate for this with thick lanugos.

Toothed whales have a two-chambered stomach similar in structure to terrestrial carnivores. They have fundic and pyloric chambers.

Locomotion

Cetaceans have two flippers on the front, and a tail fin. These flippers contain four digits. Although toothed whales do not possess fully developed hind limbs, some, such as the sperm whale, possess discrete rudimentary appendages, which may contain feet and digits. Toothed whales are fast swimmers in comparison to seals, which typically cruise at 5–15 knots, or 9–28 km/h (5.6–17.4 mph); the sperm whale, in comparison, can travel at speeds of up to 35 km/h (22 mph). The fusing of the neck vertebrae, while increasing stability when swimming at high speeds, decreases flexibility, rendering them incapable of turning their heads; river dolphins, however, have unfused neck vertebrae and can turn their heads. When swimming, toothed whales rely on their tail fins to propel them through the water. Flipper movement is continuous. They swim by moving their tail fin and lower body up and down, propelling themselves through vertical movement, while their flippers are mainly used for steering. Some species log out of the water, which may allow then to travel faster. Their skeletal anatomy allows them to be fast swimmers. Most species have a dorsal fin.

Most toothed whales are adapted for diving to great depths, porpoises are one exception. In addition to their streamlined bodies, they can slow their heart rate to conserve oxygen; blood is rerouted from tissue tolerant of water pressure to the heart and brain among other organs; haemoglobin and myoglobin store oxygen in body tissue; and they have twice the concentration of myoglobin than haemoglobin. Before going on long dives, many toothed whales exhibit a behaviour known as sounding; they stay close to the surface for a series of short, shallow dives while building their oxygen reserves, and then make a sounding dive.

Senses

Toothed whale eyes are relatively small for their size, yet they do retain a good degree of eyesight. As well as this, the eyes are placed on the sides of its head, so their vision consists of two fields, rather than a binocular view as humans have. When a beluga surfaces, its lenses and corneas correct the nearsightedness that results from the refraction of light; they contain both rod and cone cells, meaning they can see in both dim and bright light. They do, however, lack short wave-length-sensitive visual pigments in their cone cells, indicating a more limited capacity for colour vision than most mammals. Most toothed whales have slightly flattened eyeballs, enlarged pupils (which shrink as they surface to prevent damage), slightly flattened corneas, and a tapetum lucidum; these adaptations allow for large amounts of light to pass through the eye, and, therefore, a very clear image of the surrounding area. In water, a whale can see around 10.7 m (35 ft) ahead of itself, but they have a smaller range above water. They also have glands on the eyelids and outer corneal layer that act as protection for the cornea.

The olfactory lobes are absent in toothed whales, and unlike baleen whales, they lack the vomeronasal organ, suggesting they have no sense of smell.

Toothed whales are not thought to have a good sense of taste, as their taste buds are atrophied or missing altogether. However, some dolphins have preferences between different kinds of fish, indicating some sort of attachment to taste.

Sonar

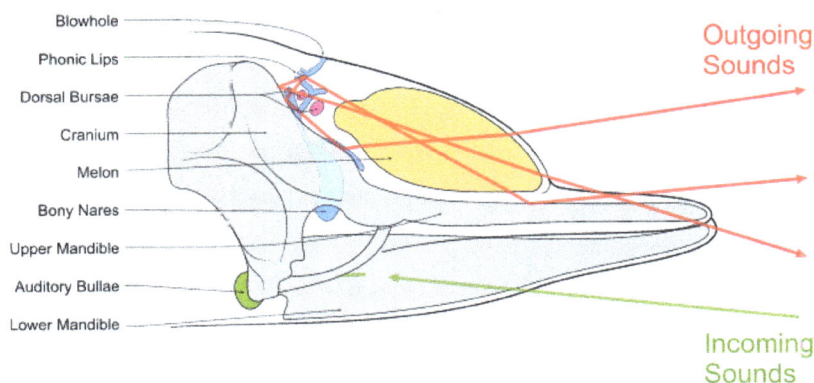

Toothed whales are capable of making a broad range of sounds using nasal airsacs located just below the blowhole. Roughly three categories of sounds can be identified: frequency-modulated whistles, burst-pulsed sounds, and clicks. Dolphins communicate with whistle-like sounds produced by vibrating connective tissue, similar to the way human vocal cords function, and through burst-pulsed sounds, though the nature and extent of that ability is not known. The clicks are directional and are used for echolocation, often occurring in a short series called a click train. The click rate increases when approaching an object of interest. Toothed whale biosonar clicks are amongst the loudest sounds made by marine animals.

The cetacean ear has specific adaptations to the marine environment. In humans, the middle ear works as an impedance equalizer between the outside air's low impedance and the cochlear fluid's high impedance. In whales, and other marine mammals, no great difference exists between the outer and inner environments. Instead of sound passing through the outer ear to the middle ear, whales receive sound through the throat, from which it passes through a low-impedance, fat-filled cavity to the inner ear. The ear is acoustically isolated from the skull by air-filled sinus pockets, which allow for greater directional hearing underwater. Odontocetes send out high-frequency clicks from an organ known as a melon. This melon consists of fat, and the skull of any such creature containing a melon will have a large depression. The melon size varies between species, the bigger it is, the more dependent they are on it. A beaked whale, for example, has a small bulge sitting on top of its skull, whereas a sperm whale's head is filled mainly with the melon.

Bottlenose dolphins have been found to have signature whistles unique to a specific individual. These whistles are used for dolphins to communicate with one another by identifying an individual. It can be seen as the dolphin equivalent of a name for humans. Because dolphins are generally associated in groups, communication is necessary. Signal masking is when other similar sounds (conspecific sounds) interfere with the original acoustic sound. In larger groups, individual whistle sounds are less prominent. Dolphins tend to travel in pods, in which the groups of dolphins range from two to 1000.

Life History and Behaviour

Intelligence

Pacific white-sided dolphins porpoising

Cetaceans are known to teach, learn, cooperate, scheme, and grieve. The neocortex of many species of dolphins is home to elongated spindle neurons that, prior to 2007, were known only in hominids. In humans, these cells are involved in social conduct, emotions, judgement, and theory of mind. Dolphin spindle neurons are found in areas of the brain homologous to where they are found in humans, suggesting they perform a similar function.

Brain size was previously considered a major indicator of the intelligence of an animal. Since most of the brain is used for maintaining bodily functions, greater ratios of brain to body mass may increase the amount of brain mass available for more complex cognitive tasks. Allometric analysis indicates that mammalian brain size scales around the two-thirds or three-quarters exponent of the body mass. Comparison of a particular animal's brain size with the expected brain size based on such allometric analysis provides an encephalisation quotient that can be used as another indication of animal intelligence. Sperm whales have the largest brain mass of any animal on earth, averaging 8,000 cm³ (490 in³) and 7.8 kg (17 lb) in mature males, in comparison to the average human brain which averages 1,450 cm³ (88 in³) in mature males. The brain to body mass ratio in some odontocetes, such as belugas and narwhals, is second only to humans.

Dolphins are known to engage in complex play behaviour, which includes such things as producing stable underwater toroidal air-core vortex rings or "bubble rings". Two main methods of bubble ring production are: rapid puffing of a burst of air into the water and allowing it to rise to the surface, forming a ring, or swimming repeatedly in a circle and then stopping to inject air into the helical vortex currents thus formed. They also appear to enjoy biting the vortex rings, so that they burst into many separate bubbles and then rise quickly to the surface. Dolphins are known to use this method during hunting. Dolphins have also been known to use tools. In Shark Bay, a population of Indo-Pacific bottlenose dolphins put sponges on their beak to protect them from abrasions and sting ray barbs while foraging in the seafloor. This behaviour is passed on from mother to daughter, and it is only observed in 54 female individuals.

Self-awareness is seen, by some, to be a sign of highly developed, abstract thinking. Self-awareness, though not well-defined scientifically, is believed to be the precursor to more advanced processes like metacognitive reasoning (thinking about thinking) that are typical of humans. Research in this field has suggested that cetaceans, among others, possess self-awareness. The most widely used test for self-awareness in animals is the mirror test, in which a temporary dye is placed on an animal's body, and the animal is then presented with a mirror; then whether the animal shows signs of self-recognition is determined. In 1995, Marten and Psarakos used television to test dolphin self-awareness. They showed dolphins real-time footage of themselves, recorded footage, and another dolphin. They concluded that their evidence suggested self-awareness rather than social behavior. While this particular study has not been repeated since then, dolphins have since "passed" the mirror test.

Vocalisations

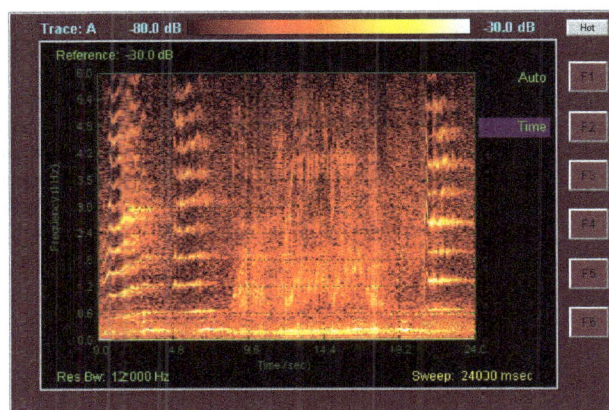

Spectrogram of dolphin vocalizations. Whistles, whines, and clicks are visible as upside down V's, horizontal striations, and vertical lines, respectively.

Dolphins are capable of making a broad range of sounds using nasal airsacs located just below the blowhole. Roughly three categories of sounds can be identified: frequency modulated whistles, burst-pulsed sounds and clicks. Dolphins communicate with whistle-like sounds produced by vibrating connective tissue, similar to the way human vocal cords function, and through burst-pulsed sounds, though the nature and extent of that ability is not known. The clicks are directional and are for echolocation, often occurring in a short series called a click train. The click rate increases when approaching an object of interest. Dolphin echolocation clicks are amongst the loudest sounds made by marine animals.

Bottlenose dolphins have been found to have signature whistles, a whistle that is unique to a specific individual. These whistles are used in order for dolphins to communicate with one another by identifying an individual. It can be seen as the dolphin equivalent of a name for humans. These signature whistles are developed during a dolphin's first year; it continues to maintain the same sound throughout its lifetime. In order to obtain each individual whistle sound, dolphins undergo vocal production learning. This consists of an experience with other dolphins that modifies the signal structure of an existing whistle sound. An auditory experience influences the whistle development of each dolphin. Dolphins are able to communicate to one another by addressing another dolphin through mimicking their whistle. The signature whistle of a male bottlenose dolphin tends to be similar to that of his mother, while the signature whistle of a female bottlenose dolphin tends to be more identifying. Bottlenose dolphins have a strong memory when it comes to these signature whistles, as they are able to relate to a signature whistle of an individual they have not encountered for over twenty years. Research done on signature whistle usage by other dolphin species is relatively limited. The research on other species done so far has yielded varied outcomes and inconclusive results.

Sperm whales can produce three specific vocalisations: creaks, codas, and slow clicks. A creak is a rapid series of high-frequency clicks that sounds somewhat like a creaky door hinge. It is typically used when homing in on prey. A coda is a short pattern of 3 to 20 clicks that is used in social situations to identify one another (like a signature whistle), but it is still unknown whether sperm whales possess individually specific coda repertoires or whether individuals make codas at different rates. Slow clicks are heard only in the presence of males (it is not certain whether females occasionally make them). Males make a lot of slow clicks in breeding grounds (74% of the time), both near the surface and at depth, which suggests they are primarily mating signals. Outside breeding grounds, slow clicks are rarely heard, and usually near the surface.

Characteristics of sperm whale clicks								
Apparent source level (dB re 1µPa [Rms])		Directionality	Centroid frequency (kHz)	Inter-click interval (s)	Duration of click (ms)	Duration of pulse (ms)	Range audible to sperm whale (km)	Inferred function	Audio sample
Usual	230	High	15	0.5–1.0	15–30	0.1	16	searching for prey	
Creak	205	High	15	0.005–0.1	0.1–5	0.1	6	homing in on prey	
Coda	180	Low	5	0.1–0.5	35	0.5	~2	social communication	
Slow	190	Low	0.5	5–8	30	5	60	communication by males	

Foraging and Predation

All whales are carnivorous and predatory. Odontocetes, as a whole, mostly feed on fish and cepha-lopods, and then followed by crustaceans and bivalves. All species are generalist and opportunistic feeders. Some may forage with other kinds of animals, such as other species of whales or certain species of pinnipeds. One common feeding method is herding, where a pod squeezes a school of fish into a small volume, known as a bait ball. Individual members then take turns plowing through the ball, feeding on the stunned fish. Coralling is a method where dolphins chase fish into shallow water to catch them more easily. Killer whales and bottlenose dolphins have also been known to drive their prey onto a beach to feed on it, a behaviour known as beach or strand feeding. The shape of the snout may correlate with tooth number and thus feeding mechanisms. The nar-whal, with its blunt snout and reduced dentition, relies on suction feeding.

Sperm whales usually dive between 300 to 800 metres (980 to 2,620 ft), and sometimes 1 to 2 ki-lometres (3,300 to 6,600 ft), in search of food. Such dives can last more than an hour. They feed on several species, notably the giant squid, but also the colossal squid, octopuses, and fish like demer-sal rays, but their diet is mainly medium-sized squid. Some prey may be taken accidentally while eating other items. A study in the Galápagos found that squid from the genera *Histioteuthis* (62%), *Ancistrocheirus* (16%), and *Octopoteuthis* (7%) weighing between 12 and 650 grams (0.026 and 1.433 lb) were the most commonly taken. Battles between sperm whales and giant squid or colos-sal squid have never been observed by humans; however, white scars are believed to be caused by the large squid. A 2010 study suggests that female sperm whales may collaborate when hunting Humboldt squid.

The killer whale is known to prey on numerous other toothed whale species. One example is the false killer whale. To subdue and kill whales, orcas continuously ram them with their heads; this can sometimes kill bowhead whales, or severely injure them. Other times, they corral their prey before striking. They are typically hunted by groups of 10 or fewer killer whales, but they are sel-dom attacked by an individual. Calves are more commonly taken by killer whales, but adults can be targeted, as well. Groups even attack larger cetaceans such as minke whales, gray whales, and rarely sperm whales or blue whales. Other marine mammal prey species include nearly 20 species of seal, sea lion and fur seal.

These cetaceans are targeted by terrestrial and pagophilic predators. The polar bear is well-adapt-ed for hunting Arctic whales and calves. Bears are known to use sit-and-wait tactics, as well as active stalking and pursuit of prey on ice or water. Whales lessen the chance of predation by gathering in groups. This, however, means less room around the breathing hole as the ice slowly closes the gap. When out at sea, whales dive out of the reach of surface-hunting killer whales. Polar bear attacks on belugas and narwhals are usually successful in winter, but rarely inflict any damage in summer.

For most of the smaller species of dolphins, only a few of the larger sharks, such as the bull shark, dusky shark, tiger shark, and great white shark, are a potential risk, especially for calves. Dolphins can tolerate and recover from extreme injuries (including shark bites) although the exact methods used to achieve this are not known. The healing process is rapid and even very deep wounds do not cause dolphins to hemorrhage to death. Even gaping wounds restore in such a way that the animal's body shape is restored, and infection of such large wounds are rare.

Lifecycle

Toothed whales are fully aquatic creatures, which means their birth and courtship behaviours are very different from terrestrial and semiaquatic creatures. Since they are unable to go onto land to calve, they deliver their young with the fetus positioned for tail-first delivery. This prevents the calf from drowning either upon or during delivery. To feed the newborn, toothed whales, being aquatic, must squirt the milk into the mouth of the calf. Being mammals, they have mammary glands used for nursing calves; they are weaned around 11 months of age. This milk contains high amounts of fat which is meant to hasten the development of blubber; it contains so much fat, it has the consistency of toothpaste. Females deliver a single calf, with gestation lasting about a year, dependency until one to two years, and maturity around seven to 10 years, all varying between the species. This mode of reproduction produces few offspring, but increases the survival probability of each one. Females, referred to as "cows", carry the responsibility of childcare, as males, referred to as "bulls", play no part in raising calves.

Interaction with Humans

Threats

Sperm Whaling

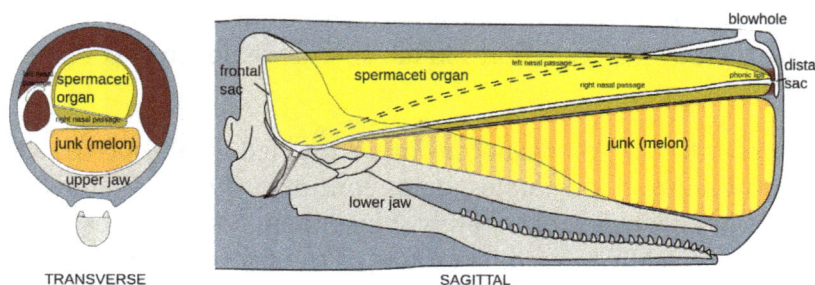

The nose of the whale is filled with a waxy substance that was widely used in candles, oil lamps, and lubricants

The head of the sperm whale is filled with a waxy liquid called spermaceti. This liquid can be refined into spermaceti wax and sperm oil. These were much sought after by 18th-, 19th-, and 20th-century whalers. These substances found a variety of commercial applications, such as candles, soap, cosmetics, machine oil, other specialized lubricants, lamp oil, pencils, crayons, leather waterproofing, rustproofing materials, and many pharmaceutical compounds. Ambergris, a solid, waxy, flammable substance produced in the digestive system of sperm whales, was also sought as a fixative in perfumery.

Sperm whaling in the 18th century began with small sloops carrying only a pair of whaleboats (sometimes only one). As the scope and size of the fleet increased, so did the rig of the vessels change, as brigs, schooners, and finally ships and barks were introduced. In the 19th-century stubby, square-rigged ships (and later barks) dominated the fleet, being sent to the Pacific (the first being the British whaleship *Emilia*, in 1788), the Indian Ocean (1780s), and as far away as the Japan grounds (1820) and the coast of Arabia (1820s), as well as Australia (1790s) and New Zealand (1790s).

Hunting for sperm whales during this period was a notoriously dangerous affair for the crews of the 19th-century whaleboats. Although a properly harpooned sperm whale generally exhibited a fairly consistent pattern of attempting to flee underwater to the point of exhaustion (at which point it would surface and offer no further resistance), it was not uncommon for bull whales to become enraged and turn to attack pursuing whaleboats on the surface, particularly if it had already been wounded by repeated harpooning attempts. A commonly reported tactic was for the whale to invert itself and violently thrash the surface of the water with its fluke, flipping and crushing nearby boats.

The estimated historic worldwide sperm whale population numbered 1,100,000 before commercial sperm whaling began in the early 18th century. By 1880, it had declined an estimated 29%. From that date until 1946, the population appears to have recovered somewhat as whaling pressure lessened, but after the Second World War, with the industry's focus again on sperm whales, the population declined even further to only 33%. In the 19th century, between 184,000 and 236,000 sperm whales were estimated to have been killed by the various whaling nations, while in the modern era, at least 770,000 were taken, the majority between 1946 and 1980. Remaining sperm whale populations are large enough so that the species' conservation status is vulnerable, rather than endangered. However, the recovery from the whaling years is a slow process, particularly in the South Pacific, where the toll on males of breeding age was severe.

Drive Hunting

Dolphins and porpoises are hunted in an activity known as dolphin drive hunting. This is accomplished by driving a pod together with boats and usually into a bay or onto a beach. Their escape is prevented by closing off the route to the ocean with other boats or nets. Dolphins are hunted this way in several places around the world, including the Solomon Islands, the Faroe Islands, Peru, and Japan, the most well-known practitioner of this method. By numbers, dolphins are mostly hunted for their meat, though some end up in dolphinariums. Despite the controversial nature of the hunt resulting in international criticism, and the possible health risk that the often polluted meat causes, thousands of dolphins are caught in drive hunts each year.

In Japan, the hunting is done by a select group of fishermen. When a pod of dolphins has been spotted, they are driven into a bay by the fishermen while banging on metal rods in the water to scare and confuse the dolphins. When the dolphins are in the bay, it is quickly closed off with nets so the dolphins cannot escape. The dolphins are usually not caught and killed immediately, but instead left to calm down over night. The following day, the dolphins are caught one by one and killed. The killing of the animals used to be done by slitting their throats, but the Japanese government banned this method, and now dolphins may officially only be killed by driving a metal pin into the neck of the dolphin, which causes them to die within seconds according to a memo from Senzo Uchida, the executive secretary of the Japan Cetacean Conference on Zoological Gardens and Aquariums. A veterinary team's analysis of a 2011 video footage of Japanese hunters killing striped dolphins using this method suggested that, in one case, death took over four minutes.

Since much of the criticism is the result of photos and videos taken during the hunt and slaughter, it is now common for the final capture and slaughter to take place on site inside a tent or under a plastic cover, out of sight from the public. The most circulated footage is probably that of the drive and subsequent capture and slaughter process taken in Futo, Japan, in October 1999, shot by the Japanese animal welfare organization Elsa Nature Conservancy. Part of this footage was, amongst others, shown on CNN. In recent years, the video has also become widespread on the internet and was featured in the animal welfare documentary *Earthlings*, though the method of killing dolphins as shown in this video is now officially banned. In 2009, a critical documentary on the hunts in Japan titled *The Cove* was released and shown amongst others at the Sundance Film Festival.

Other Threats

Toothed whales can also be threatened by humans more indirectly. They are unintentionally caught in fishing nets by commercial fisheries as bycatch and accidentally swallow fishing hooks. Gillnetting and Seine netting are significant causes of mortality in cetaceans and other marine mammals. Porpoises are commonly entangled in fishing nets. Whales are also affected by marine pollution. High levels of organic chemicals accumulate in these animals since they are high in the food chain. They have large reserves of blubber, more so for toothed whales, as they are higher up the food chain than baleen whales. Lactating mothers can pass the toxins on to their young. These pollutants can cause gastrointestinal cancers and greater vulnerability to infectious diseases. They can also be poisoned by swallowing litter, such as plastic bags. Pollution of the Yangtze river has led to the extinction of the Baiji. Environmentalists speculate that advanced naval sonar endangers some whales. Some scientists suggest that sonar may trigger whale beachings, and they point to signs that such whales have experienced decompression sickness.

Conservation

Currently, no international convention gives universal coverage to all small whales, although the International Whaling Commission has attempted to extend its jurisdiction over them. ASCOBANS was negotiated to protect all small whales in the North and Baltic Seas and in the northeast Atlantic. ACCOBAMS protects all whales in the Mediterranean and Black Seas. The global UNEP Convention on Migratory Species currently covers seven toothed whale species or populations on its Appendix I, and 37 species or populations on Appendix II. All oceanic cetaceans are listed in CITES appendices, meaning international trade in them and products derived from them is very limited.

Numerous organisation are dedicated to protecting certain species that do not fall under any international treaty, such as the Committee for the Recovery of the Vaquita, and the Wuhan Institute of Hydrobiology (for the Yangtze finless porpoise).

Species

Various species of toothed whales, mainly dolphins, are kept in captivity, as well as several other species of porpoise such as harbour porpoises and finless porpoises. These small cetaceans are more often than not kept in theme parks, such as SeaWorld, commonly known as a dolphinarium. Bottlenose dolphins are the most common species kept in dolphinariums, as they are relatively easy to train, have a long lifespan in captivity, and have a friendly appearance. Hundreds if not thousands of Bottlenose Dolphins live in captivity across the world, though exact numbers are hard to determine. Killer whales are well known for their performances in shows, but the number kept in captivity is very small, especially when compared to the number of bottlenose dolphins, with only 44 captives being held in aquaria as of 2012. Other species kept in captivity are spotted Dolphins, false killer whales, and common dolphins, Commerson's dolphins, as well as rough-toothed dolphins, but all in much lower numbers than the bottlenose dolphin. Also, fewer than ten pilot whales, Amazon river dolphins, Risso's dolphins, spinner dolphins, or tucuxi are in captivity.

Two unusual and very rare hybrid dolphins, known as wolphins, are kept at the Sea Life Park in Hawaii, which is a cross between a bottlenose dolphin and a false killer whale. Also, two common/bottlenose hybrids reside in captivity: one at Discovery Cove and the other at SeaWorld San Diego.

Controversy

Organisations such as Animal Welfare Institute and the Whale and Dolphin Conservation Society campaign against the captivity of dolphins and killer whales. SeaWorld faced a lot of criticism after the documentary *Blackfish* was released in 2013.

Aggression among captive killer whales is common. In August 1989, a dominant female killer whale, Kandu V, attempted to rake a newcomer whale, Corky II, with her mouth during a live show, and smashed her head into a wall. Kandu V broke her jaw, which severed an artery, and then bled to death. In November 2006, a dominant female killer whale, Kasatka, repeatedly dragged experienced trainer Ken Peters to the bottom of the stadium pool during a show after hearing her calf crying for her in the back pools. In February 2010, an experienced female trainer at SeaWorld Orlando, Dawn Brancheau, was killed by killer whale Tilikum shortly after a show in Shamu Stadium. Tilikum had been associated with the deaths of two people previously. In May 2012, Occupational Safety and Health Administration administrative law judge Ken Welsch cited SeaWorld for two violations in the death of Dawn Brancheau and fined the company a total of US$12,000. Trainers were banned from making close contact with the killer whales. In April 2014, the US Court of Appeals for the District of Columbia denied an appeal by SeaWorld.

In 2013, SeaWorld's treatment of killer whales in captivity was the basis of the movie *Blackfish*, which documents the history of Tilikum, a killer whale captured by SeaLand of the Pacific, later transported to SeaWorld Orlando, which has been involved in the deaths of three people. In the aftermath of the release of the film, Martina McBride, 38 Special, REO Speedwagon, Cheap Trick, Heart, Trisha Yearwood, and Willie Nelson cancelled scheduled concerts at SeaWorld parks. SeaWorld disputes the accuracy of the film, and in December 2013 released an ad countering the allegations and emphasizing its contributions to the study of cetaceans and their conservation.

Baleen Whale

Baleen whales (systematic name Mysticeti), known earlier as whalebone whales, form a parvorder of the infraorder Cetacea (whales, dolphins and porpoises). They are a widely distributed and diverse parvorder of carnivorous marine mammals. Mysticeti comprise the families Balaenidae (right whales), Balaenopteridae (rorquals), Cetotheriidae (the pygmy right whale), and Eschrichtiidae (the gray whale). There are currently 15 species of baleen whale. While cetaceans were historically thought to have descended from mesonychids, molecular evidence supports them as relatives of even-toed ungulates (Artiodactyla). Baleen whales split from toothed whales (Odontoceti) around 34 million years ago.

Baleen whales range in size from the 20 ft (6 m) and 6,600 lb (3,000 kg) pygmy right whale to the 112 ft (34 m) and 190 t (210 short tons) blue whale, which is also the largest creature on earth. They are sexually dimorphic. Baleen whales can have streamlined or large bodies, depending on the feeding behavior, and two limbs that are modified into flippers. Though not as flexible and agile as seals, baleen whales can swim very fast, with the fastest able to travel at 23 miles per hour (37 km/h). Baleen whales use their baleen plates to filter out food from the water by either lunge-feeding or skim-feeding. Baleen whales have fused neck vertebrae, and are unable to turn their head at all. Baleen whales have two blowholes. Some species are well adapted for diving to great depths. They have a layer of fat, or blubber, under the skin to keep warm in the cold water.

Although baleen whales are widespread, most species prefer the colder waters around the Northern and Southern poles. Gray whales are specialized for feeding on bottom-dwelling crustaceans. Rorquals are specialized at lunge-feeding, and have a streamlined body to reduce drag while accelerating. Right whales skim-feed, meaning they use their enlarged head to effectively take in a large amount of water and sieve the slow-moving prey. Males typically mate with more than one female (polygyny), although the degree of polygyny varies with the species. Male strategies for reproductive success vary between performing ritual displays (whale song) or lek mating. Calves are typically born in the winter and spring months and females bear all the responsibility for raising them. Mothers fast for a relatively long period of time over the period of migration, which varies between species. Baleen whales produce a number of vocalizations, notably the songs of the humpback whale.

The meat, blubber, baleen, and oil of baleen whales have traditionally been used by the indigenous peoples of the Arctic. Once relentlessly hunted by commercial industries for these products, cetaceans are now protected by international law. However, the North Atlantic right whale is ranked critically endangered by the International Union for Conservation of Nature. Besides hunting, baleen whales also face threats from marine pollution and ocean acidification. It has been speculated that man-made sonar results in strandings. They have rarely been kept in captivity, and this has only been attempted with juveniles or members of one of the smallest species.

Taxonomy

Baleen whales are cetaceans classified under the parvorder Mysticeti, and consist of four extant families: Balaenidae (right whales), Balaenopteridae (rorquals), Cetotheriidae (pygmy right whale), and Eschrichtiidae (gray whale). Balaenids are distinguished by their enlarged head and thick blubber, while rorquals and gray whales generally have a flat head, long throat pleats, and are more streamlined than Balaenids. Rorquals also tend to be longer than the latter. Cetaceans (whales, dolphins, and porpoises) and artiodactyls are now classified under the order Cetartiodactyla, often still referred to as Artiodactyla (given that the cetaceans are deeply nested with the artiodactyls). The closest living relatives to baleen whales are toothed whales both from the infraorder Cetacea.

Balaenidae consists of two genera: *Eubalaena* (right whales) and *Balaena* (the bowhead whale, *B. mysticetus*). Balaenidae was thought to have consisted of only one genus until studies done through the early 2000s reported that bowhead whales and right whales are morphologically (different skull shape) and phylogenically different. According to a study done by H. C. Rosenbaum (of the American Museum of Natural History) and colleagues, the North Pacific (*E. japonica*) and Southern right (*E. australis*) whales are more closely related to each other than to the North Atlantic right whale (*E. glacialis*).

Rorquals consist of two genera (*Balaenoptera* and *Megaptera*) and nine species: the fin whale (*B. physalus*), the Sei whale (*B. borealis*), Bryde's whale (*B. brydei*), Eden's whale (*B. edeni*), the blue whale (*B. musculus*), the common minke whale (*B. acutorostrata*), the Antarctic minke whale (*B. bonaerensis*), Omura's whale (*B. omurai*), and the humpback whale (*M. novaeangliae*). In a 2012 review of cetacean taxonomy, Alexandre Hassanin (of the Muséum National d'Histoire Naturelle) and colleagues suggested that, based on phylogenic criteria, there are four extant genera of rorquals. They recommend that the genus *Balaenoptera* be limited to the fin whale, have minke whales fall under the genus *Pterobalaena*, and have *Rorqualus* contain the Sei whale, Bryde's whale, Eden's whale, the blue whale, and Omura's whale.

Cetotheriidae consists of only one living member: the pygmy right whale (*Caperea marginata*). The first descriptions date back to the 1840s of bones and baleen plates resembling a smaller version of the right whale, and was named *Balaena marginata*. In 1864, it was moved into the genus *Caperea* after a skull of another specimen was discovered. Six years later, the pygmy right whale was classified under the family Neobalaenidae. Despite its name, the pygmy right whale is more genetically similar to rorquals and gray whales than to right whales. A study published in 2012, based on bone structure, moved the pygmy right whale from the family Neobalaenidae to the family Cetotheriidae, making it a living fossil; Neobalaenidae was elevated down to subfamily level as Neobalaeninae.

Eschrichtiidae consists of only one living member: the gray whale (*Eschrichtius robustus*). The two populations, one in the Sea of Okhotsk and Sea of Japan and the other in the Mediterranean Sea and East Atlantic, are thought to be genetically and physiologically dissimilar. The gray whale is traditionally placed as the only living species in its genus and family. However, DNA analysis by studies, such as the one by Takeshi Sasaki (of the Tokyo Institute of Technology) and colleagues, indicates certain rorquals, such as the humpback whale, *Megaptera novaeangliae*, and the fin whale, *Balaenoptera physalus*, are more closely related to the gray whale than they are to some other rorquals, such as the minke whale, *Balaenoptera acutorostrata*.

Etymology

Mysticetes are also known as baleen whales because of the presence of baleen. These animals rely on their baleen plates to sieve plankton and other small organisms from the water. The term "baleen" (Middle English *baleyn, ballayne, ballien, bellane,* etc.) is an archaic word for "whale", derived from the Latin word *balæna*.

Right whales got their name because of whalers preferring them over other species; they were essentially the "right whale" to catch.

Differences between Families

Rorquals use throat pleats to expand their mouth which allows them to feed more effectively. However, rorquals need to build up water pressure in order to expand their mouth, leading to a lunge-feeding behavior. Lunge-feeding is where a whale rams a bait ball (a swarm of small fish) at high speeds. Rorquals generally have a streamlined physique to reduce drag in the water while doing this. Balaenids rely on their huge head, as opposed to the rorquals' throat pleats, to feed effectively. This feeding behavior allows them to grow very big and bulky, without the necessity for a streamlined body. They have callosities, unlike other whales, with the exception of the bowhead whale. Rorquals have a higher proportion of muscle tissue and tend to be negatively buoyant, whereas right whales have a higher proportion of blubber and are positively buoyant. The gray whale is easily distinguished from other extant cetaceans by its sleet-gray color, a dorsal ridge (knuckles on the back), and its gray-white scars left from parasites. Like in the rorquals, their throat pleats increase the capacity of their throat, allowing them to filter larger volumes of water at once. Gray whales are bottom-feeders, meaning they sift through sand to get their food. They usually turn on their side and scoop up sediment into their mouth and filter out benthic creatures like amphipods, which leaves a noticeable mark on their head. The pygmy right whale is easily confused with minke whales because of their similar characteristics, such as its small size, dark gray top, light gray bottom, and a light eye-patch.

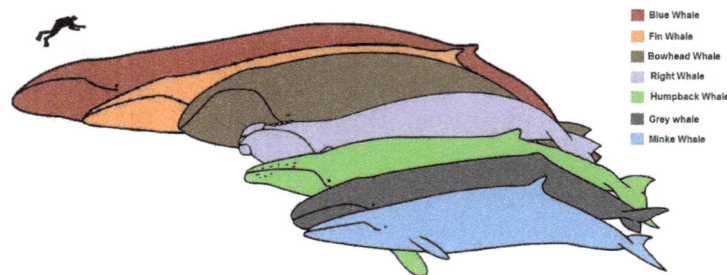

Baleen whales vary considerably in size and shape, depending on their feeding behavior
(note the whale in blue is actually a sei whale)

The four mysticete families		
Eschrichtiidae	*Eubalaena*, Balaenidae *Balaena*, Balaenidae	*Balaenoptera*, Balaenopteridae
Megaptera, Balaenopteridae	Cetotheriidae	

List of Mysticetes

- The "†" signs denote extinct families and genera.

- Parvorder Mysticeti: baleen whales

    - Family †Aetiocetidae

- †*Aetiocetus*
- †*Ashcrocetus*
- †*Chonecetus*
- †*Fucaia*
- †*Morawanocetus*
- †*Willungacetus*
  - †Family Llanocetidae
    - †*Llanocetus*
  - †Family Mammalodontidae
    - †*Janjucetus*
    - †*Mammalodon*
- Clade Chaeomysticeti
  - Superfamily Eomysticetoidea
    - †Family Cetotheriopsidae
      - †*Cetotheriopsis*
    - †Family Eomysticetidae
      - †*Eomysticetus*
      - †*Micromysticetus*
      - †*Tohoraata*
      - †*Tokarahia*
      - †*Waharoa*
      - †*Yamatocetus*
  - Clade Balaenomorpha
    - Superfamily Balaenoidea
      - Family Balaenidae: right whales and bowhead whale
        - *Balaena* – bowhead whales
        - †*Balaenella*
        - †*Balaenotus*
        - †*Balaenula*
        - *Eubalaena* – right whales
        - †*Idiocetus*

- †*Morenocetus*
- †*Peripolocetus*
- Clade Thalassotherii
  - †*Hibacetus*
  - †*Isocetus*
  - Family Cetotheriidae
    - †*Brandtocetus*
    - †*Cephalotropis*
    - †*Cetotherium*
    - †*Eucetotherium*
    - †*Herentalia*
    - †*Herpetocetus*
    - †*Joumocetus*
    - †*Kurdalagonus*
    - †*Metopocetus*
    - †*Nannocetus*
    - †*Otradnocetus*
    - †*Palaeobalaena?*
    - †*Piscobalaena*
    - †*Titanocetus?*
    - †*Vampalus*
    - †*Zygiocetus*
  - Subfamily Neobalaeninae
    - *Caperea*, pygmy right whale
    - †*Miocaperea*
- Superfamily Balaenopteroidea
  - †*Eobalaenoptera*
  - †*Mauicetus*
  - †*Tiphyocetus*
  - Family †Aglaocetidae

- † *Aglaocetus*
- † *Isanacetus*
- † *Pinocetus*
- Family Balaenopteridae: rorquals
  - †*Archaebalaenoptera*
  - *Balaenoptera*
  - †*Burtinopsis* (*nomen dubium*)
  - †*Cetotheriophanes*
  - †*Diunatans*
  - *Megaptera* – humpback whale
  - †*Notiocetus*
  - †*Parabalaenoptera*
  - †*Plesiobalaenoptera*
  - †*Plesiocetus*
  - †*Praemegaptera*
  - †*Protororqualus*
- †Family Diorocetidae
  - †*Amphicetus*
  - †*Diorocetus*
  - †*Plesiocetopsis*
  - †*Thinocetus*
  - †*Uranocetus*
- Subfamily Eschrichtiinae
  - †*Archaeschrichtius*
  - †*Eschrichtioides*
  - *Eschrichtius* – gray whales

Gray whale skeleton

- †*Gricetoides*
- †*Megapteropsis* (*nomen dubium*)
- †Family Pelocetidae (invalid subgroup)
  - †*Cophocetus*
  - †*Halicetus*
  - †*Parietobalaena*
  - †*Pelocetus*
- †Family Tranatocetidae
  - †*Mesocetus*
  - †*Mixocetus*
  - †*Tranatocetus*
- *incertae sedis*
  - *Amphiptera* (existence unconfirmed)
  - †*Horopeta*
  - †*Imerocetus*
  - †*Mioceta* (*nomen dubium*)
  - †*Piscocetus*
  - †*Siphonocetus* (*nomen dubium*)
  - †*Tretulias* (*nomen dubium*)
  - †*Ulias* (*nomen dubium*)

Evolutionary History

Restoration of *Janjucetus hunderi*

Mysticeti split from Odontoceti (toothed whales) 26 to 17 million years ago during the Eocene. Their evolutionary link to archaic toothed cetaceans (Archaeoceti) remained unknown until the extinct *Janjucetus hunderi* was discovered in the early 1990s in Victoria, Australia. Like a modern baleen whale, *Janjucetus* had baleen in its jaw and had very little biosonar capabilities. However, its jaw also contained teeth, with incisors and canines built for stabbing and molars and premolars built for tearing. These early mysticetes were exceedingly small compared to modern baleen whales, with species like *Mammalodon* measuring no greater than 10 feet (3 m). It is thought that their size increased with their dependence on baleen. The discovery of *Janjucetus* and others like it suggests that baleen evolution went through several transitional phases. Species like *Mammalodon colliveri* had little to no baleen, while later species like *Aetiocetus weltoni* had both baleen and teeth, suggesting they had limited filter feeding capabilities; later genera like *Cetotherium* had no teeth in their mouth, meaning they were fully dependent on baleen and could only filter feed.

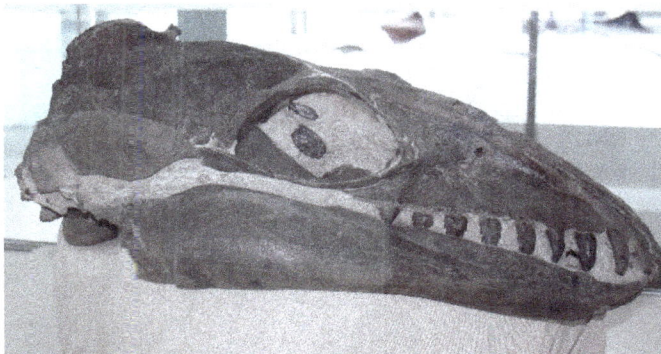

Archaeomysticetes, like *Janjucetus*, had teeth.

Fucaia buelli is the earliest mysticete, dating back to 33 million years ago (mya). Measuring only 6.6 feet (2 m), it is the smallest baleen whale discovered. It is only known from its teeth; they suggest a suction feeding behavior, much like that of beaked whales. Like other early toothed mysticetes, or "archaeomysticetes", *F. buelli* had heterodont dentition. Archaeomysticetes from the Oligocene are the Mammalodontidae (*Mammalodon* and *Janjucetus*) from Australia. They were small with shortened rostra, and a primitive dental formula (3.1.4.33.1.4.3). In baleen whales, enlarged mouths adapted for suction feeding evolved before specializations for bulk filter feeding. In the toothed Oligocene mammalodontid *Janjucetus*, the symphysis is short and the mouth enlarged, the rostrum is wide, and the edges of the maxillae are thin, indicating an adaptation for suction feeding. The aetiocetid *Chonecetus* still had teeth, but the presence of a groove on the interior side of each mandible indicates the symphysis was elastic, which would have enabled rotation of each mandible, an initial adaptation for bulk feeding like in modern mysticetes.

The lineages of rorquals and right whales split almost 20 mya. It is unknown where this occurred, but it is generally believed that they, like their descendants, followed plankton migrations. These primitive mysticetes had lost their heterodont dentition in favor of baleen, and are believed to have lived on a specialized benthic, plankton, or copepod diet like modern mysticetes. Mysticetes experienced their first radiation in the mid-Miocene. Balaenopterids grew bigger during this time, with species like *Balaenoptera sibbaldina* rivaling the blue whale in terms of size. It is thought this radiation was caused by global climate change and major tectonic activity (the Antarctic Circumpolar Current).

The first toothless ancestors of Mysticetes appeared before the first radiation in the late Oligocene. *Eomysticetus* and others like it showed no evidence in the skull of echolocation abilities, suggesting they mainly relied on their eyesight for navigation. The eomysticetes had long, flat rostra that lacked teeth and had external nares located halfway up the dorsal side of the snout. Though the palate is not well-preserved in these specimens, they are thought to have had baleen and been filter feeders. Miocene baleen whales were preyed upon by larger predators like killer sperm whales and *Megalodon*.

Miocene mysticetes were often hunted by megalodon and killer sperm whales.

Anatomy

A humpback whale skeleton. Notice how the jaw is split into two.

Motion

When swimming, baleen whales rely on their flippers for locomotion in a wing-like manner similar to penguins and sea turtles. Flipper movement is continuous. While doing this, baleen whales use their tail fluke to propel themselves forward through vertical motion while using their flippers for steering, much like an otter. Some species leap out of the water, which may allow them to travel faster. Because of their great size, right whales are not flexible or agile like dolphins, and none can move their neck because of the fused cervical vertebrae; this sacrifices speed for stability in the water. The hind legs are enclosed inside the body, and are thought to be vestigial organs. However, a 2014 study suggests that the pelvic bone serves as support for whale genitalia.

Rorquals, needing to build speed to feed, have several adaptions for reducing drag, including a streamlined body; a small dorsal fin, relative to its size; and lack of external ears or long hair. The fin whale, the fastest among baleen whales, can travel at 23 miles per hour (37 km/h). While feeding, the rorqual jaw expands to a volume that can be bigger than the whale itself; to do this, the mouth inflates. The inflation of the mouth causes the cavum ventrale, the throat pleats on the underside stretching to the navel, to expand, increasing the amount of water that the mouth can store. The mandible is connected to the skull by dense fibers and cartilage (fibrocartilage), allowing the jaw to swing open at almost a 90° angle. The mandibular symphysis is also fibrocartilaginous, allowing the jaw to bend which lets in more water. To prevent stretching the mouth too far, rorquals have a sensory organ located in the middle of the jaw to regulate these functions.

External Anatomy

Paired blowholes of a humpback and the V-shaped blow of a right whale

Baleen whales have two flippers on the front, near the head. Like all mammals, baleen whales breathe air and must surface periodically to do so. Their nostrils, or blowholes, are situated at the top of the cranium. Baleen whales have two blowholes, as opposed to toothed whales which have one. These paired blowholes are longitudinal slits that converge anteriorly and widen posteriorly, which causes a V-shaped blow. They are surrounded by a fleshy ridge that keeps water away while the whale breathes. The septum that separates the blowholes has two plugs attached to it, making the blowholes water-tight while the whale dives.

Like other mammals, the skin of baleen whales has an epidermis, a dermis, a hypodermis, and connective tissue. The epidermis, the pigmented layer, is 0.2 inches (5 mm) thick, along with connective tissue. The epidermis itself is only 0.04 inches (1 mm) thick. The dermis, the layer underneath the epidermis, is also thin. The hypodermis, containing blubber, is the thickest part of the skin and functions as a means to conserve heat. Right whales have the thickest hypodermis of any cetacean, averaging 20 inches (51 cm), though, as in all whales, it is thinner around openings (such as the blowhole) and limbs. Blubber may also be used to store energy during times of fasting. The connective tissue between the hypodermis and muscles allows only limited movement to occur between them. Unlike in toothed whales, baleen whales have small hairs on the top of their head, stretching from the tip of the rostrum to the blowhole, and, in right whales, on the chin. Like other marine mammals, they lack sebaceous and sweat glands.

The baleen of baleen whales are keratinous plates. They are made of a calcified hard α-keratin material, a fiber-reinforced structure made of intermediate filaments (proteins). The degree of calcification varies between species, with the sei whale having 14.5% hydroxyapatite, a mineral that coats teeth and bones, whereas minke whales have 1–4% hydroxyapatite. In most mammals, keratin structures, such as wool, air-dry, but aquatic whales rely on calcium salts to form on the plates to stiffen them. Baleen plates are attached to the upper jaw and are absent in the mid-jaw, forming two separate combs of baleen. The plates decrease in size as they go further back into the jaw; the largest ones are called the "main baleen plates" and the smallest ones are called the "accessory plates". Accessory plates taper off into small hairs.

Accessory baleen plates taper off into small hairs

Unlike other whales (and most other mammals), the females are larger than the males. Sexual dimorphism is usually reversed, with the males being larger, but the females of all baleen whales are usually five percent larger than males. Sexual dimorphism is also displayed through whale song, notably in humpback whales where the males of the species sing elaborate songs. Male right whales have bigger callosities than female right whales. The males are generally more scarred than females which is thought to be because of aggression during mating season.

Internal Systems

The unique lungs of baleen whales are built to collapse under the pressure instead of resisting the pressure which would damage the lungs, enabling some, like the fin whale, to dive to a depth of −1,540 feet (−470 m). The whale lungs are very efficient at extracting oxygen from the air, usually 80%, whereas humans only extract 20% of oxygen from inhaled air. Lung volume is relatively low compared to terrestrial mammals because of the inability of the respiratory tract to hold gas while diving. Doing so may cause serious complications such as embolism. Unlike other mammals, the lungs of baleen whales lack lobes and are more sacculated. Like in humans, the left lung is smaller than the right to make room for the heart. To conserve oxygen, blood is rerouted from pressure-tolerant-tissue to internal organs, and they have a high concentration of myoglobin which allows them to hold their breath longer.

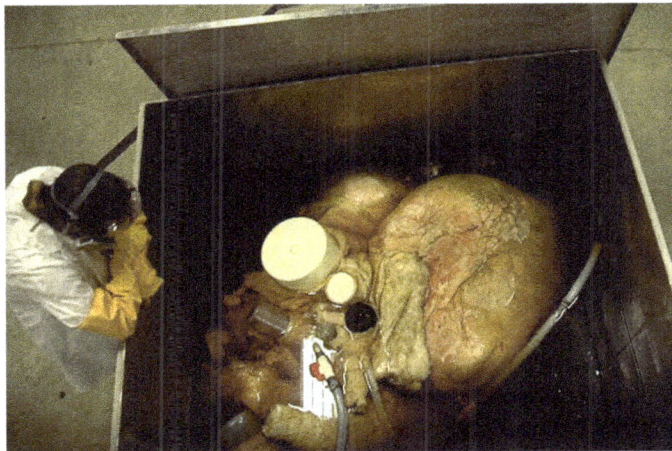

The heart of a blue whale with a person standing next to it

The heart of baleen whales functions similarly to other mammals, with the major difference being the size. The heart can reach 1,000 pounds (454 kg), but is still proportional to the whale's size. The muscular wall of the ventricle, which is responsible for pumping blood out of the heart, can be 3 to 5 inches (7.6 to 12.7 cm) thick. The aorta, an artery, can be .75 inches (1.9 cm) thick. Their resting heart rate is 60 to 140 beats per minute (bpm), as opposed to the 60 to 100 bpm in humans. When diving, their heart rate will drop to 4 to 15 bpm to conserve oxygen. Like toothed whales, they have a dense network of blood vessels (rete mirabile) which prevents heat-loss. Like in most mammals, heat is lost in their extremities, so, in baleen whales, warm blood in the arteries is surrounded by veins to prevent heat loss during transport. As well as this, heat inevitably given off by the arteries warms blood in the surrounding veins as it travels back into the core. This is otherwise known as countercurrent exchange. To counteract overheating while in warmer waters, baleen whales reroute blood to the skin to accelerate heat-loss. They have the largest blood corpuscles (red and white blood cells) of any mammal, measuring 4.1×10^{-4} inches (10 μm) in diameter, as opposed to human's 2.8×10^{-4}-inch (7.1 μm) blood corpuscles.

When sieved from the water, food is swallowed and travels through the esophagus where it enters a three-chambered-stomach. The first compartment is known as the fore-stomach; this is where food gets ground up into an acidic liquid, which is then squirted into the main stomach. Like in humans, the food is mixed with hydrochloric acid and protein-digesting enzymes. Then, the partly digested food is moved into the third stomach, where it meets fat-digesting enzymes, and is then mixed with an alkaline liquid to neutralize the acid from the fore-stomach to prevent damage to the intestinal tract. Their intestinal tract is highly adapted to absorb the most nutrients from food; the walls are folded and contain copious blood vessels, allowing for a greater surface area over which digested food and water can be absorbed. Baleen whales get the water they need from their food; however, the salt content of most of their prey (invertebrates) are similar to that of seawater, whereas the salt content of a whale's blood is considerably lower (three times lower) than that of seawater. The whale kidney is adapted to excreting excess salt; however, while producing urine more concentrated than seawater, it wastes a lot of water which must be replaced.

Baleen whales have a relatively small brain compared to their body mass. Like other mammals, their brain has a large, folded cerebrum, the part of the brain responsible for memory and processing sensory information. Their cerebrum only makes up about 68% of their brain's weight, as opposed to human's 83%. The cerebellum, the part of the brain responsible for balance and coordination, makes up 18% of their brain's weight, compared to 10% in humans, which is probably due to the great degree of control necessary for constantly swimming. Necropsies on the brains of gray whales revealed iron oxide particles, which may allow them to find magnetic north like a compass.

Unlike most animals, whales are conscious breathers. All mammals sleep, but whales cannot afford to become unconscious for long because they may drown. They are believed to exhibit unihemispheric slow-wave sleep, in which they sleep with half of the brain while the other half remains active. This behavior was only documented in toothed whales until footage of a humpback whale sleeping (vertically) was shot in 2014.

It is largely unknown how baleen whales produce sound because of the lack of a melon and vocal cords. In a 2007 study, it was discovered that the larynx had U-shaped folds which are thought to be similar to vocal cords. They are positioned parallel to air flow, as opposed to the perpendicular vocal cords of terrestrial mammals. These may control air flow and cause vibrations. The walls of the larynx are able to contract which may generate sound with support from the arytenoid cartilages. The muscles surrounding the larynx may expel air rapidly or maintain a constant volume while diving.

Senses

Their eyes are relatively small for their size.

The eyes of baleen whales are relatively small for their size and are positioned near the end of the mouth. This is probably because they feed on slow or immobile prey, combined with the fact that most sunlight does not pass 30 feet (9.1 m), and hence they do not need acute vision. A whale's eye is adapted for seeing both in the euphotic and aphotic zones by increasing or decreasing the pupil's size to prevent damage to the eye. As opposed to land mammals which have a flattened lens, whales have a spherical lens. The retina is surrounded by a reflective layer of cells (tapetum lucidum), which bounces light back at the retina, enhancing eyesight in dark areas. However, light is bent more near the surface of the eye when in air as opposed to water; consequently, they can see much better in the air than in the water. The eyeballs are protected by a thick outer layer to prevent abrasions, and an oily fluid (instead of tears) on the surface of the eye. Baleen whales appear to have limited color vision, as they lack S-cones.

The mysticete ear is adapted for hearing underwater, where it can hear sound frequencies as low as 7 Hz and as high as 22 kHz. It is largely unknown how sound is received by baleen whales. Unlike in toothed whales, sound does not pass through the lower jaw. The auditory meatus is blocked by connective tissue and an ear plug, which connects to the eardrum. The inner-ear bones are contained in the tympanic bulla, a bony capsule. However, this is attached to the skull, suggesting that vibrations passing through the bone is important. Sinuses may reflect vibrations towards the cochlea. It is known that when the fluid inside the cochlea is disturbed by vibrations, it triggers sensory hairs which send electrical current to the brain, where vibrations are processed into sound.

Baleen whales have a small, yet functional, vomeronasal organ. This allows baleen whales to detect chemicals and pheromones released by their prey. It is thought that 'tasting' the water is important for finding prey, and track down other whales. They are believed to have an impaired sense of smell due to the lack of the olfactory bulb, but they do have an olfactory tract. Baleen whales have few if any taste buds, suggesting they have lost their sense of taste. They do retain salt-receptor taste-buds suggesting that they can taste saltiness.

Behavior

Migration

It is thought that plankton blooms dictate where whales migrate. This usually occurs in the polar regions during the sunny spring and summer months, bringing along other plankton such as euphausiids which whales feed on. They also migrate to calving grounds in tropical waters during the winter months when plankton populations are low. As well as this, newborns, with underdeveloped blubber, would likely die of frostbite in the winter temperatures. It is also postulated by a 2008 study that these take place to avoid calves being predated on by killer whales. The migration cycle is repeated annually. The gray whale has the longest recorded migration of any mammal, with one traveling 14,000 miles (23,000 km) from the Sea of Okhotsk to the Baja Peninsula.

Foraging

All baleen whales are carnivorous; however a 2015 study revealed they house gut flora similar to that of terrestrial herbivores. Different kinds of prey are found in different abundances depending on location, and each type of whale is adapted to a specialized way of foraging. There are two types of feeding behaviors: skim-feeding and lunge-feeding, but some species do both depending on the type and amount of food. For example, Antarctic whales mostly feed on euphausiids; however, this is mainly effective for lunge-feeders, whereas skim-feeders, like the right whales, feed primarily on copepods. They feed alone or in small groups. Baleen whales get the water they need from their food, and their kidneys excrete excess salt.

The lunge-feeders are the rorquals. To feed, lunge-feeders expand the volume of their jaw to a volume bigger than the original volume of the whale itself; to do this, the mouth inflates to expand the mouth. The inflation of the mouth causes the throat pleats to expand, increasing the amount of water that the mouth can store. Just before they ram the baitball, the jaw swings open at almost a 90° angle and bends which lets in more water. To prevent stretching the mouth too far, rorquals have a sensory organ located in the middle of the jaw to regulate these functions. Then they must decelerate. This process takes a lot of mechanical work, and is only energy-effective when used against a large baitball. The skim-feeders, are right whales, gray whales, pygmy right whales, and sei whales (which also lunge feed). To feed, skim-feeders swim with an open mouth, filling it with water and prey. Prey must occur in sufficient numbers to trigger the whale's interest, be within a certain size range so that the baleen plates can filter it, and be slow enough so that it cannot escape. The "skimming" may take place on the surface, underwater, or even at the ocean's bottom, indicated by mud occasionally observed on right whales' bodies. Gray whales feed primarily on the ocean's bottom, feeding on benthic creatures.

Predation and Parasitism

Baleen whales, primarily juveniles and calves, are preyed on by killer whales. It is thought that annual whale migration occurs to protect the calves from the killer whales. There have also been reports of a pod of killer whales attacking and killing an adult bowhead whale, by holding down its flippers, covering the blowhole, and ramming and biting until death. Generally, a mother and calf pair, when faced with the threat of a killer whale pod, will either fight or flee. Fleeing only occurs in species that can swim away quickly, the rorquals. Slower whales must fight the pod alone or with a small family group. There has been one report of a shark attacking and killing a whale calf. This occurred in 2014 during the sardine run when a shiver of dusky sharks attacked a humpback whale calf. Usually, the only shark that will attack a whale is the cookie cutter shark, which leaves a small, non-fatal bite mark.

Orange whale lice on a right whale

Many parasites latch onto whales, notably whale lice and whale barnacles. Almost all species of whale lice are specialized towards a certain species of whale, and there can be more than one species per whale. Whale lice eat dead skin, resulting in minor wounds in the skin. Whale louse infestations are especially evident in right whales, where colonies propagate on their callosities. Though not a parasite, whale barnacles latch onto the skin of a whale during their larval stage. However, in doing so it does not harm nor benefit the whale, so their relationship is often labeled as an example of commensalism. Some baleen whales will deliberately rub themselves on substrate to dislodge parasites. Some species of barnacle, such as *Conchoderma auritum* and whale barnacles, attach to the baleen plates, though this seldom occurs. A species of copepod, *Balaenophilus unisetus*, inhabits baleen plates of whales in tropical waters. A species of Antarctic diatom, *Cocconeis ceticola*, forms a film on the skin, which takes a month to develop; this film causes minor damage to the skin. They are also plagued by internal parasites such as stomach worms, cestodes, nematodes, liver flukes, and acanthocephalans.

Reproduction and Development

Female right whale with calf

Before reaching adulthood, baleen whales grow at an extraordinary rate. In the blue whale, the largest species, the fetus grows by some 220 lb (100 kg) per day just before delivery, and by 180 lb (80 kg) per day during suckling. Before weaning, the calf increases its body weight by 17 t (17 long tons; 19 short tons) and grows from 23 to 26 ft (7 to 8 m) at birth to 43 to 52 ft (13 to 16 m) long. When it reaches sexual maturity after 5–10 years, it will be 66 to 79 ft (20 to 24 m) long and possibly live as long as 80–90 years. Calves are born precocial, needing to be able to swim to the surface at the moment of their birth.

Most rorquals mate in warm waters in winter to give birth almost a year later. A 7-to-11 month lactation period is normally followed by a year of rest before mating starts again. Adults normally start reproducing when 5–10 years old and reach their full length after 20–30 years. In the smallest rorqual, the minke whale, 10 ft (3 m) calves are born after a 10-month pregnancy and weaning lasts until it has reached about 16 to 18 ft (5 to 5.5 m) after 6–7 months. Unusual for a baleen whale, female minkes (and humpbacks) can become pregnant immediately after giving birth; in most species, there is a two-to-three-year calving period. In right whales, the calving interval is usually three years. They grow very rapidly during their first year, after which they hardly increase in size for several years. They reach sexual maturity when 43 to 46 ft (13 to 14 m) long. Baleen whales are K-strategists, meaning they raise one calf at a time, have a long life-expectancy, and a low infant mortality rate. Some 19th century harpoons found in harvested bowheads indicate this species can live more than 100 years. Baleen whales are promiscuous, with none showing pair bonds. They are polygynous, in that a male may mate with more than one female. The scars on male whales suggest they fight for the right to mate with females during breeding season, somewhat similar to lek mating.

Baleen whales have fibroelastic (connective tissue) penises, similar to those of artiodactyls. The tip of the penis, which tapers toward the end, is called the *pars intrapraeputialis* or *terminal cone*. The blue whale has the largest penis of any organism on the planet, typically measuring 8–10 feet (2.4–3.0 m). Accurate measurements of the blue whale are difficult to take because the whale's erect length can only be observed during mating. The penis on a right whale can be up to 2.7 m (8.9 ft) – the testes, at up to 2 m (6.6 ft) in length, 78 cm (2.56 ft) in diameter, and weighing up to 525 lb (238 kg), are also the largest of any animal on Earth.

Whale Song

All baleen whales use sound for communication and are known to "sing", especially during the breeding season. Blue whales produce the loudest sustained sounds of any animals: their low-frequency (about 20 Hz) moans can last for half a minute, reach almost 190 decibels, and be heard hundreds of kilometers away. Adult male humpbacks produce the longest and most complex songs; sequences of moans, groans, roars, sighs, and chirps sometimes lasting more than ten minutes are repeated for hours. Typically, all humpback males in a population sing the same song over a breeding season, but the songs change slightly between seasons, and males in one population have been observed adapting the song from males of a neighboring population over a few breeding seasons.

Intelligence

Unlike their toothed whale counterparts, baleen whales are hard to study because of their immense size. Intelligence tests such as the mirror test cannot be done because their bulk and lack of body language makes a reaction impossible to be definitive. However, studies on the brains of humpback whales revealed spindle cells, which, in humans, control theory of mind. Because of this, it is thought that baleen whales, or at least humpback whales, have consciousness.

Relationship with Humans

History of Whaling

Whaling by humans has existed since the Stone Age. Ancient whalers used harpoons to spear the bigger animals from boats out at sea. People from Norway started hunting whales around 4,000 years ago, and people from Japan began hunting whales in the Pacific at least as early as that. Whales are typically hunted for their meat and blubber by aboriginal groups; they used baleen for baskets or roofing, and made tools and masks out of bones. The Inuit hunted whales in the Arctic Ocean. The Basques started whaling as early as the 11th century, sailing as far as Newfoundland in the 16th century in search of right whales. 18th and 19th century whalers hunted down whales mainly for their oil, which was used as lamp fuel and a lubricant, and baleen (or whalebone), which was used for items such as corsets and skirt hoops. The most successful whaling nations at this time were the Netherlands, Japan, and the United States.

Commercial whaling was historically important as an industry well throughout the 19th and 20th centuries. Whaling was at that time a sizable European industry with ships from Britain, France, Spain, Denmark, the Netherlands, and Germany, sometimes collaborating to hunt whales in the Arctic. By the early 1790s, whalers, namely the British (Australian) and Americans, started to focus efforts in the South Pacific; in the mid 1900s, over 50,000 humpback whale were taken from the South Pacific. At its height in the 1880s, U.S. profits turned to USD10,000,000, equivalent to USD225,000,000 today. Commonly exploited species included arctic whales such as the gray whale, right whale, and bowhead whale because they were close to the main whaling ports, like New Bedford. After those stocks were depleted, rorquals in the South Pacific were targeted by nearly all whaling organizations; however, they often out-swam whaling vessels. Whaling rorquals was not effective until the harpoon cannon was invented in the late 1860s. Whaling basically stopped when stocks of all species were depleted to a point that they could not be harvested on a commercial scale. Whaling was controlled in

1982 when the International Whaling Commission (IWC) placed a moratorium setting catch limits to protect species from dying out from over-exploitation, and eventually banned it:

Notwithstanding the other provisions of paragraph 10, catch limits for the killing for commercial purposes of whales from all stocks for the 1986 coastal and the 1985/86 pelagic seasons and thereafter shall be zero. This provision will be kept under review, based upon the best scientific advice, and by 1990 at the latest the Commission will undertake a comprehensive assessment of the effects of this decision on whale stocks and consider modification of this provision and the establishment of other catch limits.

–IWC Commission Schedule, paragraph 10(e)

Conservation and Management Issues

As of 2013, the International Union for Conservation of Nature (IUCN) recognizes 15 mysticete species. One species—the North Atlantic right whale—is Endangered with only around 400(±50) individuals left, and four more are also classified as Endangered (North Pacific right whale, the blue whale, the fin whale, and the Sei whale), and another 5 ranked as Data deficient (Bryde's whale, Eden's whale, Omura's whale, Southern minke whale, and pygmy right whale). Species that live in polar habitats are vulnerable to the effects of ongoing climate change, particularly declines in sea ice, as well as ocean acidification.

The whale watching industry and anti-whaling advocates argue that whaling catches "friendly" whales that are curious about boats, as these whales are the easiest to catch. This analysis claims that once the economic benefits of hotels, restaurants and other tourist amenities are considered, hunting whales is a net economic loss. This argument is particularly contentious in Iceland, as it has among the most-developed whale-watching operations in the world and the hunting of minke whales resumed in August 2003. Brazil, Argentina and South Africa argue that whale watching is a growing billion-dollar industry that provides more revenue than commercial whaling would provide. Peru, Uruguay, Australia, and New Zealand also support proposals to permanently forbid whaling south of the Equator, as Solor (an island of Indonesia) is the only place of the Southern Hemisphere that takes whales. Anti-whaling groups, such as the International Fund for Animal Welfare (IFAW), claim that countries which support a pro-whaling stance are damaging their economies by driving away anti-whaling tourists.

Protests against Japan's scientific whaling

Commercial whaling was historically important for the world economy. All species were exploited, and as one type's stock depleted, another type was targeted. The scale of whale harvesting decreased substantially through the 1960s as all whale stocks had been depleted, and practically stopped in 1988 after the International Whaling Commission placed a moratorium which banned whaling for commercial use. Several species that were commercially exploited have rebounded in numbers; for example, gray whales may be as numerous as they were prior to whaling, making it the first marine mammal to be taken off the Endangered species list. The Southern right whale was hunted to near extinction in the mid-to-late 20th century, with only a small (unknown) population around Antarctica. Because of international protection, the Southern right whale's population has been growing 7% annually since 1970. Conversely, the eastern stock of North Atlantic right whale was extirpated from much of its former range, which stretched from the coast of North Africa to the North Sea and Iceland; it is thought that the entire stock consists of only ten individuals, making the eastern stock functionally extinct.

Baleen whales continue to be harvested. However, only three nations take whales: Iceland, Norway, and Japan. All these nations are part of the IWC, with Norway and Iceland rejecting the moratorium and continuing commercial whaling. Japan, being part of the IWC, whales under the Scientific Permit stated in Article VIII in the Convention for the Regulation of Whaling, which allows the taking of whales for scientific research. Japan has had two main research programs: the Joint Aquatic Resources Permit Application (JARPA) and the Japanese Research Program in the North (JARPN). JARPN is focused in the North Pacific and JARPA around the Antarctic. JARPA mainly caught Antarctic minke whales, catching nearly 7,000; to a far lesser extent, they also caught fin whales. Animal-rights activist groups, such as the Greenpeace, object to Japan's scientific whaling, with some calling it a substitute for commercial whaling. In 2014, the International Court of Justice (the UN judicial branch) banned the taking of whales for any purpose in the Southern Ocean Whale Sanctuary; however, Japan refuses to stop whaling and has only promised to cut their annual catches by a third (around 300 whales per year).

The remains of a North Atlantic right whale after it collided with a ship propeller.

Baleen whales can also be affected by humans in more indirect ways. For species like the North Atlantic right whale, which migrates through some of the world's busiest shipping lanes, the biggest threat is from being struck by ships. The Lloyd's mirror effect results in low frequency propeller sounds not being discernible near the surface, where most accidents occur. Combined with spreading and acoustic shadowing effects, the result is that the whale is unable to hear an approaching vessel before it has been run over or entrapped by the hydrodynamic forces of the vessel's passage. A 2014 study noted that a lower vessel speed correlated with lower collision rates. The ever-increasing amount of ocean noise, including sonar, drowns out the vocalizations produced by whales, notably in the blue whale which produces the loudest vocalization, which makes it harder for them to communicate. Blue whales stop producing foraging D calls once a mid-frequency sonar is activated, even though the sonar frequency range (1–8 kHz) far exceeds their sound production range (25–100 Hz). Poisoning from toxic substances such as Polychlorinated biphenyl (PCB) is generally low because of their low trophic level. Some baleen whales can become victims of bycatch, which is especially serious for North Atlantic right whales considering there are only 450 left. Right whales feed with a wide-open mouth, risking entanglement in any rope or net fixed in the water column. Rope wraps around their upper jaw, flippers and tail. Some are able to escape, but others remain entangled. If observers notice, they can be successfully disentangled, but others die over a period of months. Other whales, such as humpback whales, can also be entangled.

In Captivity

A gray whale in captivity

Baleen whales have rarely been kept in captivity. Their large size and appetite make them expensive creatures to maintain. Pools of proper size would also be very expensive to build. For example, a single gray whale calf would need to eat 475 pounds (215 kg) of fish per day, and the pool would have to accommodate the 13-foot (4 m) calf, along with ample room to swim. Only gray whales have survived being kept in captivity for over a year. The first gray whale, which was captured in Scammon's Lagoon, Baja California Sur, in 1965, was named Gigi and died two months later from an infection. The second gray whale, which was captured in 1971 from the same lagoon, was named Gigi II and was released a year later after becoming too big. The last gray whale, J.J., beached itself in Marina del Rey, California, where it was rushed to SeaWorld San Diego and, after 14 months, was released because it got too big to take care of. Reaching 19,200 pounds (8,700 kg) and 31 feet (9.4 m), J.J. was the largest creature to be kept in captivity. The Mito Aquarium in Numazu, Shizuoka, Japan, housed three minke whales in the nearby bay enclosed by nets. One survived for three months, another (a calf) survived for two weeks, and another was kept for over a month before breaking through the nets.

Balaenidae

Balaenidae is a family of whales of the parvorder Mysticeti that contains two living genera. Historically, it is known as the right whale family, as it was thought to contain only species of right whales.

Through most of the 20th Century, however, that became a much-debated (and unresolved) topic amongst the scientific community. Finally, in the early 2000s, science reached a definitive conclusion: the bowhead whale, once commonly known as the Greenland right whale, was not in fact a right whale. The family of Balaenids, therefore, comprises the right whales (genus *Eubalaena*), and in a genus all to its own, the very closely related bowhead whale (genus *Balaena*).

Evolutionary History

Baleen whales belong to the monophyletic lineage of Mysticeti order. Mysticeti are large filter-feeding cetaceans that also included of largest animals on earth as well as some of the most critically endangered. Based on morphology and molecular data, four extant family-level clades are recognized within Mysticeti: Balaenidae (bowhead and right whales), Neobalaenidae (pygmy right whales), Eschirichtiidae (gray whales), and Balaenopteridae (rorquals). Phylogenetic relationships of the mysticeti order remain unclear due to legal and logistical challenges. However, recent morphological analysis, support Balaenidae as a monophyletic group that is the sister group to Neobalaenidae.

Characteristics

Balaenids are large whales, with an average adult length of 15 to 17 metres (45–50 feet), and weighing 50-80 tonnes. Their principle distinguishing feature is their narrow, arched, upper jaw, which gives the animals a deeply curved jawline. This shape allows for especially long baleen plates. The animals utilise these by floating at or near the surface, and straining food from the water, which they then scrape off the baleen with their tongues - a feeding method that contrasts with those of the rorquals and the gray whale. Their diet consists of small crustaceans, primarily copepods, although some species also eat a significant amount of krill.

Balaenids are also robustly built by comparison with the rorquals, and lack the grooves along the throat that are distinctive of those animals. They have exceptionally large heads in comparison with their bodies, reaching 40% of the total length in the case of the Bowhead Whale. They have short, broad, flippers, and lack a dorsal fin.

All species are at least somewhat migratory, moving into warmer waters during the winter, during which they both mate and give birth. Gestation lasts 10–11 months, results in the birth of a single young, and typically occurs once every three years.

Distribution

The four species of the Balaenidae that are found throughout temperate and polar waters; Eubalaena glacialis (North Atlantic right whale), Eubalaena japonica (North Pacific right whale), Eubalaena australis (Southern right whale), and Balaena mysticetus (Bowhead whale). Bowhead and right whales can reach up to 18 meters in length and over 100 tons at maturity.

Exploitation and Conservation Status

Members of Balaenidae can live over 70 years and were hunted extensively in the late 1800s for their blubber. Approximately 40% of right whales body mass is blubber and thus were known as the "right" whale to kill. After death, the large blubber deposits caused right whales to float to the surface, which facilitated an easier oil harvest. With population estimated between 300 -350, the North Atlantic right whale is the most critically endangered great whale. The Northern Pacific right whale is also endangered with only about 500 individuals extant. The Southern right whale (~7500 individuals in 1997) and the Bowhead whale (20,000 to 40,000) have made stronger recoveries since whale hunting was significantly curtailed by international agreement.

Taxonomy

- Family Balaenidae
  - Genus *Balaena*
    - Bowhead whale, *Balaena mysticetus*
  - Genus *Eubalaena*
    - North Atlantic right whale, *Eubalaena glacialis*
    - North Pacific right whale, *Eubalaena japonica*
    - Southern right whale, *Eubalaena australis*

Until recently, all right whales of the genus *Eubalaena* were considered a single species—*E. glacialis*. In 2000, genetic studies of right whales from the different ocean basins led scientists to conclude that the populations in the North Atlantic, North Pacific and Southern Hemisphere constitute three distinct species. Further genetic analysis in 2005 using mitochondrial DNA and nuclear DNA has supported the conclusion that the three populations should be treated as separate species, and the separation has been adopted for management purposes by the U.S. National Marine Fisheries Service and the International Whaling Commission.

The cladogram is a tool for visualizing and comparing the evolutionary relationships between taxa. The point where a node branches off is analogous to an evolutionary branching – the diagram can be read left-to-right, much like a timeline. The following cladogram of the family Balaenidae serves to illustrate the current scientific consensus as to the relationships between the North Pacific right whale and the other members of its family.

Family Balaenidae		
Family Balaenidae	*Eubalaena* (right whales)	*E. glacialis* North Atlantic right whale
		E. japonica North Pacific right whale
		E. australis Southern right whale
	Balaena (bowhead whales)	*B. mysticetus* bowhead whale

Rorqual

Rorquals (Balaenopteridae) are the largest group of baleen whales, a family with nine extant species in two genera. They include what is believed to be the largest animal that has ever lived, the blue whale, which can reach 180 tonnes (200 short tons), and the fin whale, which reaches 120 tonnes (130 short tons); even the smallest of the group, the northern minke whale, reaches 9 tonnes (9.9 short tons).

Rorquals take their name from French *rorqual*, which derives from the Norwegian word *røyrkval*, meaning "furrow whale".

Characteristics

All members of the family have a series of longitudinal folds of skin running from below the mouth back to the navel (except the sei whale and common minke whale, which have shorter grooves). These are understood to allow the mouth to expand immensely when feeding, "permitting them to engorge great mouthfuls of food and water in a single gulp". These "pleated throat grooves" distinguish balaenopterids from other whales.

Rorquals are slender and streamlined in shape, compared with their relatives the right whales, and most have narrow, elongated flippers. They have a dorsal fin, situated about two-thirds the way back. Rorquals feed by gulping in water, and then pushing it out through the baleen plates with their tongue. They feed on crustaceans, such as krill, but also on various fish, such as herrings and sardines.

Gestation in rorquals lasts 11–12 months, so that both mating and birthing occur at the same time of year. Cows give birth to a single calf, which is weaned after 6–12 months, depending on species. Of some species, adults live in small groups, or "pods" of two to five individuals. For example, humpback whales have a fluid social structure, often engaging behavioral practices in a pod, other times being solitary.

The "minke" whale is allegedly named after a Norwegian whaler named Meincke, who mistook a northern minke whale for a blue whale.

Distribution and Habitat

Distribution is worldwide: the blue, fin, humpback, and the sei whales are found in all major oceans; the common (northern) and Antarctic (southern) minke whale species are found in all the oceans of their respective hemispheres; and either of Bryde's whale and Eden's whale occur in the Atlantic, Pacific, and Indian oceans, being absent only from the cold waters of the Arctic and Antarctic.

Most rorquals are strictly oceanic: the exceptions are Bryde's whale and Eden's whale (which are usually found close to shore all year round) and the humpback whale (which is oceanic but passes close to shore when migrating). It is the largest and the smallest types — the blue whale and Antarctic minke whale — that occupy the coldest waters in the extreme south; the fin whale tends not to approach so close to the ice shelf; the sei whale tends to stay further north again. (In the northern hemisphere, where the continents distort weather patterns and ocean currents, these movements are less obvious, although still present.) Within each species, the largest individuals tend to approach the poles more closely, while the youngest and fittest ones tend to stay in warmer waters before leaving on their annual migration.

Most rorquals breed in tropical waters during the winter, then migrate back to the polar feeding grounds rich in plankton and krill for the short polar summer.

Feeding Habits

Humpback feeding on young pollock off Alaska

As well as other methods, rorqual whales obtain prey by lunge feeding on bait balls. Lunge feeding is an extreme feeding method, where the whale accelerates to a high velocity and then opens its mouth to a large gape angle. This generates the water pressure required to expand its mouth and engulf and filter a huge amount of water and fish.

Rorquals have a number of anatomical features that enable them to do this, including bilaterally separate mandibles, throat pleats that can expand to huge size, and a unique sensory organ consisting of a bundle of mechanoreceptors that helps their brains to coordinate the engulfment action. Furthermore, their large nerves are flexible so that they can stretch and recoil. In fact, they give rorquals the ability to open their mouths so wide that they would be capable of taking in water at volumes greater than their own sizes. These nerves are packed into a central core area that is surrounded by elastin fibers. Opening the mouth causes the nerves to unfold, and they snap back after the mouth is closed. According to Potvin and Goldbogen, lunge feeding in rorqual whales represents the largest biomechanical event on Earth.

Taxonomy

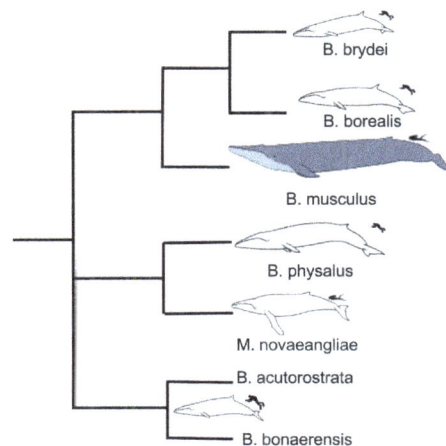

Cladogram of the family Balaenopteridae using complete mtDNA sequences and short interspersed repetitive element (SINE) insertion data.

Skeleton of the extinct *"Megaptera" hubachi* at the Museum für Naturkunde, Berlin

Formerly, the rorqual family Balaenopteridae was split into two subfamilies, the Balaenopterinae and the Megapterinae, with each subfamily contained one genus, *Balaenoptera* and *Megaptera*, respectively. However, the phylogeny of the various rorqual species shows the current division is paraphyletic, and in 2005, the division into subfamilies was dropped. The discovery of a new species of balaenopterid, Omura's whale (*Balaenoptera omurai*), was announced in November 2003, which looks similar to, but smaller than, the fin whale; individuals of this species were found in Indo-Pacific waters.

- Family Balaenopteridae: Rorquals
  - †*Archaebalaenoptera*
  - *Balaenoptera*
    - Fin whale, *Balaenoptera physalus*
    - Sei whale, *Balaenoptera borealis*
    - Bryde's whale, *Balaenoptera brydei*
    - Eden's whale, *Balaenoptera edeni*
    - Blue whale, *Balaenoptera musculus*
    - Common minke whale, *Balaenoptera acutorostrata*
    - Antarctic minke whale, *Balaenoptera bonaerensis*
    - Omura's whale, *Balaenoptera omurai*
  - †*Cetotheriophanes*
  - †*Diunatans*
  - †*Incakujira*
  - *Megaptera*
    - Humpback whale, *Megaptera novaeangliae*
  - †*Parabalaenoptera*
  - †*Plesiobalaenoptera*

- o †*Plesiocetus*
- o †*Praemegaptera*
- o †*Protororqualus*

Alternate Generic Taxonomy for Living Rorquals

- *Balaenoptera*
    - o Fin whale, *Balaenoptera physalus*
- *Megaptera*
    - o Humpback whale, *Megaptera novaeangliae*
- *Pterobalaena*
    - o Common minke whale, *Pterobalaena acutorostrata*
    - o Antarctic minke whale, *Pterobalaena bonaerensis*
- *Rorqualus*
    - o Sei whale, *Rorqualus borealis*
    - o Bryde's whale, *Rorqualus brydei*
    - o Eden's whale, *Rorqualus edeni*
    - o Blue whale, *Rorqualus musculus*
    - o Omura's whale, *Rorqualus omurai*

Pygmy Right Whale

The pygmy right whale (*Caperea marginata*) is a member of the cetotheres, a family of baleen whales, which until 2012 were thought to be extinct; previously *C. marginata* was considered the sole member of the family Neobalaenidae. First described by John Edward Gray in 1846, it is the smallest of the baleen whales, ranging between 6 metres (20 ft) and 6.5 metres (21 ft) in length and 3,000 and 3,500 kg in mass. Despite its name, the pygmy right whale may have more in common with the gray whale and rorquals than the bowhead and right whales.

The pygmy right whale is found in the Southern Ocean in the lower reaches of the Southern Hemisphere, and feeds on copepods and euphausiids. Little is known about its population or social habits. Unlike most other baleen whales, it has rarely been subject to exploitation.

Taxonomy

During the 1839-43 voyage of James Clark Ross, naturalists found bones and baleen plates resembling a smaller version of the right whale. In his *Zoology of the Voyage of HMS Erebus and Terror* (1846), John Edward Gray described the new species, naming it *Balaena marginata*. In 1864, Gray established a new genus (*Caperea*) after receiving a skull and some bones of another specimen. Six years later, in 1870, he added the name *Neobalaena*. He soon realized the three species were one and the same: *Caperea marginata* (*caperea* means "wrinkle" in Latin, "referring to the wrinkled appearance of the ear bone"; while *marginata* translates to "enclosed with a border", which "refers to the dark border around the baleen plates of some individuals"). In research findings published on December 18, 2012, paleontologist Felix Marx compared the skull bones of pygmy right whales to those of other extinct cetaceans, finding it to be a close relative to the Cetotheriidae, making the pygmy right whale a living fossil.

In 2012, Italian palaeontologist Michelangelo Bisconti described *Miocaperea pulchra*, a first fossil pygmy right whale from Peru. This new genus differs from the living genus in some cranial details, but Bisconti's study confirmed the monophyly of the Neobalaenidae and he concluded that the rorqual-like features in *C. marginata* must be the result of parallel evolution. The presence of a fossil neobalaenid some 2,000 km (1,200 mi) north of the known range of *C. marginata*, suggests that environmental change has caused a southern shift in neobalaenid distribution. A second, un-described species was tentatively assigned to Neobalaenidae in 2012.

Description

The pygmy right whale is rarely encountered and consequently little studied. However, it is known that it is by far the smallest of the baleen whales. Calves are estimated to be about 1.6 metres (5 ft 3 in) to 2.2 metres (7 ft 3 in) at birth (a 2 metres (6 ft 7 in) fetus was reported from a 6 metres (20 ft) female). By the time they are weaned they may be about 3 metres (9.8 ft) to 3.5 metres (11 ft) long. It is believed they become sexually mature at about 5 metres (16 ft) and physically mature at about 6 metres (20 ft). The longest male was a 6.1 metres (20 ft) individual which had stranded in Cloudy Bay, Tasmania, while the longest female was a 6.45 metres (21.2 ft) individual which had stranded in Stanley, Tasmania. Pygmy rights can weigh as much as 3,430 kg—a 6.21 metres (20.4 ft) female weighed 3,200 kg and a 5.47 metres (17.9 ft) male weighed 2,850 kg. Gestation and lactation pe-riods and longevity are all unknown. Part of the reason for the scarcity of data may be the relative inactivity of the whale, making location for study difficult. The blow is small and indistinct and the whale is usually a slow undulating swimmer, although capable of bursts of acceleration.

The coloring and shape of the pygmy right whale, dark gray dorsally and lighter gray ventrally, commonly with a pair of chevron-shaped lighter patches behind the eyes, is similar to that of the dwarf minke and Antarctic minke whales and at sea may easily be confused with these two species if the jaw and flippers are not carefully observed. The arched jawline is not as pronounced as other right whales and may not be sufficient to distinguish a pygmy right whale from a minke whale. The long, narrow cream-coloured baleen plates with a distinctive white gumline are the most effective discriminators. Unlike true right whales, pygmy rights do not have callosities. The dorsal fin is fal-cate (crescent-shaped) and located about three-quarters of the way along the back of the animal. Unlike the minke whales, occasionally the dorsal will not be seen on the whale surfacing. Like the minkes, though, it doesn't raise its flukes when it dives.

The skull and skeleton of the pygmy right whale is unlike those of any other extant whale: the supraoccipital shield extends farther posteriorly; the ear bone has a lateral wrinkle and is roughly square in outline. All seven cervical vertebrae are fused, and the pygmy right has only 44 vertebrae. The 18 pairs of ribs are broad and flat, and make up 39–45% of the vertebral column (compared to 33% in other mysticetes.) Each thoracic vertebrae has a pair of huge wing-like transverse processes, many of which overlap. The dorsal end of the ribs are remarkably thin and almost fail to make contact with the transverse processes. The reduced tail (or sacrocaudal region) features a vestigial pelvis and small chevron bones. The flippers have four digits. The lungs and heart are relatively small which suggests that the pygmy right is not a deep diver. The larynx is reported to be different from any other cetacean. Like other mysticetes, the pygmy right has a large laryngeal sac, but in contrast to other mysticetes, this sac is positioned on the right side of the midline in the pygmy right. The presence of this laryngeal sac can possibly be the explanation for the long thorax and flattened ribs, but the peculiar ribs have been the source many speculations.

Analysis of the stomach contents of dead pygmy right whales indicates that it feeds on copepods and euphausiids. The social and mating structures are unknown. The whale is typically seen alone or in pairs, sometimes associated with other cetaceans (including dolphins, pilot whales, minke whales, and once a sei whale cow and calf). Occasionally larger groups are seen — in 2001 a group of 14 were seen at 46°S in the South Pacific about 450 km southeast of New Zealand, while in 1992 a group of about 80 individuals were seen 320 nm southwest of Cape Leeuwin and another group of over 100 individuals were sighted in June 2007 about 40 km southwest of Portland, Victoria.

Population and Distribution

The pygmy right whale is among the least studied cetaceans (as of 2008 fewer than 25 "at sea" sightings of the species have been made). The species lives in the Southern Hemisphere and is believed to be circumpolar, living in a band from about 30°S to 55°S in areas with surface water temperature between 5 and 20 °C (41 and 68 °F). Individuals have been found on the coasts of Chile, Tierra del Fuego, Namibia, South Africa, Australia and New Zealand. One group may be a year-round resident off Tasmania. The total population is unknown.

Whaling and Whale-watching

On account of its relatively small size and sparse distribution the pygmy right whale has rarely been taken by whalers. A 3.39 metres (11.1 ft) male was taken off South Africa in 1917, and it is likely that a few pygmy rights were taken opportunistically by whalers hunting minke whales. Also a few pygmy right whales are known to have been caught in fishing nets. However these factors are not believed to have had a significant impact on the population.

Most data about pygmy right whales come from individual specimens washed up on coastlines; they are rarely encountered at sea and so they are not the primary subject of any whale watching cruises.

Conservation

The pygmy right whale is listed on Appendix II of the Convention on the Conservation of Migratory Species of Wild Animals (CMS). It is listed on Appendix II as it has an unfavourable conservation status or would benefit significantly from international co-operation organised by tailored agreements.

The pygmy right whale is also covered by Memorandum of Understanding for the Conservation of Cetaceans and Their Habitats in the Pacific Islands Region (Pacific Cetaceans MOU).

Eschrichtiidae

Eschrichtiidae or the gray whales is a family of baleen whale (suborder Mysticeti) with a single extant species, the gray whale (*Eschrichtius robustus*). The family, however, also includes three described fossil genera: *Archaeschrichtius* and *Eschrichtioides* from the Miocene and Pliocene of Italy respectively, and *Gricetoides* from the Pliocene of North Carolina. The names of the extant genus and the family honours Danish zoologist Daniel Eschricht.

Taxonomic History

A number of 18th century authors described the gray whale as *Balaena gibbosa*, the "whale with six bosses", apparently based on a brief note by Dudley 1725:

The Scrag Whale is near a kin to the Fin-back, but instead of a Fin upon his Back, the Ridge of the Afterpart of his Back is cragged with half a Dozen Knobs or Nuckles; he is nearest the right Whale in Figure and for Quantity of Oil; his Bone is white, but won't split.

The gray whale was first described as a distinct species by Lilljeborg 1861 based on a subfossil found in the brackish Baltic Sea, apparently a specimen from the now extinct north Atlantic population. Lilljeborg, however, identified it as "*Balaenoptera robusta*", a species of rorqual. Gray 1864 realized that the rib and scapula of the specimen was different from those of any known rorquals, and therefore erected a new genus for it, *Eschrichtius*. Van Beneden & Gervais 1868 were convinced that the bones described by Lilljeborg could not belong to a living species but that they were similar to fossils that Van Beneden had described from the harbour of Antwerp (most of his named species are now considered nomina dubia) and therefore named the gray whale *Plesiocetus robustus*, reducing Lilljeborg's and Gray's names to synonyms.

Scammon 1869 produced one of the earliest descriptions of living Pacific gray whales, and notwithstanding that he was among the whalers who nearly drove them to extinction in the lagoons of the Baja California Peninsula, they were and still are associated with him and his description of the species. At this time, however, the extinct Atlantic population was considered a separate species (*Eschrischtius robustus*) from the living Pacific population (*Rhachianectes glaucus*).

Things got increasingly confused as 19th century scientists introduced new species at an alarming rate (e.g. *Eschrichtius pusillus*, *E. expansus*, *E. priscus*, *E. mysticetoides*), often based on fragmentary specimens, and taxonomists started to use several generic and specific names interchangeably and not always correctly (e.g. *Agalephus gobbosus*, *Balaenoptera robustus*, *Agalephus gibbosus*). Things got even worse in the 1930s when it was finally realised that the extinct Atlantic population was the same species as the extant Pacific population, and the new combination *Eschrichtius gibbosus* was proposed.

In his morphological analysis, Bisconti 2008 found that eschrichtiids and Cetotheriidae (*Cetotherium*, *Mixocetus* and *Metopocetus*) form a monophyletic sister group of Balaenopteridae.

A specimen from the Late Pliocene of northern Italy, named *"Cetotherium" gastaldii* by Strobel 1875 and renamed *"Balaenoptera" gastaldii* by Portis 1885, was identified as a basal eschrichtiid by Bisconti 2008 who recombined it to *Eschrichtioides gastaldii*.

Steeman et al. 2009 found that the gray whale is phylogenetically distinct from rorquals and that previous morphological studies were correct in the conclusion that the evolution of gulp feeding was a single event in the rorqual lineage.

Evolution

Fossils of Eschrichtiidae have been found in all major oceanic basins in the Northern Hemisphere, and the family is believed date back to the Late Miocene. Today, gray whales are only present in the northern Pacific, but a population was also present in the northern Atlantic before being driven to extinction by European whalers three centuries ago.

Fossil eschrichtiids from before the Holocene are rare compared to other fossil mysticetes. The only Pleistocene fossil from the Pacific referred to *E. eschrichtius* is a partial skeleton and an associated skull from California, estimated to be about 200 thousand years old. However, a late Pliocene fossil from Hokkaido, Japan, referred to *Eschrichtius* sp. is estimated to be 2.6 to 3.9 Mya and a similar unnamed fossil has been reported from California.

In their description of *Archaeschrichtius ruggieroi* from the late Miocene of Italy, Bisconti & Varola 2006 argued that eschrichtiids most likely originated in the Mediterranean Basin about 10 million years ago and remained there, either permanently or intermittently, at least until the Early Pliocene (5–3 Mya).

Marine Otter

The marine otter (*Lontra felina*) is a rare and poorly known South American mammal of the weasel family (Mustelidae). The scientific name means "otter cat", and in Spanish, the marine otter is also often referred to as *gato marino*: "marine cat". The marine otter (while spending much of its time out of the water) only lives in saltwater, coastal environments and rarely ventures into freshwater or estuarine habitats. This saltwater exclusivity is unlike most other otter species, including the almost fully aquatic sea otter (*Enhydra lutris*) of the northeast Pacific.

Description

Marine otters are relatively small, and among otters, only the oriental small-clawed otter is smaller. However the latter species inhabits freshwater sites; thus the marine otter is the smallest exclusively marine mammal on Earth. Lengths range from 83 to 113 centimetres (33 to 44 in), not counting the tail of 30 to 36 centimetres (12 to 14 in). Weights can range from 3 to 5.8 kilograms (6.6 to 12.8 lb). Their fur is dark brown on the back and light brown on belly. The guard hairs cover short insulating fur with a grayish color. The fur is coarser and tougher than in sea otters.

The front and hind paws are webbed, and there are four teats.

The marine otter's lower jaws contain eight pairs of teeth, and the upper jaws eight or nine pairs. The teeth are developed for slicing rather than crushing.

Sexual dimorphism in this species not readily apparent.

Distribution and Habitat

Marine otters are found in littoral areas of southwestern South America, close to shore and in the intertidal areas of northern Peru (from the port of Chimbote), along the entire coast of Chile, and the extreme southern reaches of Argentina. Occasional vagrant sightings still occur as far afield as the Falkland Islands.

The marine otter mainly inhabits rocky shorelines with abundant seaweed and kelp, and infrequently visits estuaries and freshwater rivers. It appears to select habitats with surprisingly high exposure to strong swells and winds, unlike many other otters, which prefer calmer waters. Caves and crevices in the rocky shorelines may provide them with the cover they need, and often a holt will have no land access at high tide. Marine otters avoid sandy beaches.

Feeding

Little is known about the diet of marine otters, but their primary prey is believed to be crab, shrimp, mollusks, and fish.They also eat many types of crustaceans.

Behavior and Reproduction

Marine otters are most often seen individually or in small groups of up to three. They are difficult to spot, swimming low in the water, exposing only their heads and backs. It is not known whether they are territorial, as males are occasionally seen fighting, yet fights have also been observed even between mating pairs. Fighting takes place on prominent rocks above the waterline, which are also used for resting, feeding, and grooming. Marine otters have also been observed feeding cooperatively on large fish, but it is not known how common the practice is.

The otters are diurnal, primarily active in the daytime.

Marine otters may be monogamous or polygamous, and breeding occurs in December or January. Litters of two to five pups are born in January, February or March after a gestation period of 60 to 70 days. The pups remain with their mother for about 10 months of parental care, and can sometimes be seen on the mother's belly as she swims on her back, a practice similar to that of the sea otter. Parents bring food to the pups and teach them to hunt.

Conservation Status

Marine otters are rare and are protected under Peruvian, Chilean, and Argentine law. In the past, they were extensively hunted both for their fur and due to perceived competition with fisheries. Hunting extirpated them from most of Argentina and the Falkland Islands. Poaching is still a problem, but one of unknown magnitude. It is unknown how many marine otters exist in the wild or what habitats should be preserved to encourage their recovery. Marine otters were listed under CITES Appendix I in 1976, and are listed as endangered by the U.S. Department of the Interior.

References

- Grandin, Temple; Johnson, Catherine (2009). "Wildlife". Animals Make Us Human: Creating the Best Life for Animals. p. 240. ISBN 978-0-15-101489-7.

- Thewissen, J. G. M.; Perrin, William R.; Wirsig, Bernd (2002). "Hearing". Encyclopedia of Marine Mammals. San Diego: Academic Press. pp. 570–572. ISBN 978-0-12-551340-1.

- Thomas, Jeanette A.; Kastelein, Ronald A. (1990). Sensory Abilities of Cetaceans: Laboratory and Field Evidence. 196. New York: Springer Science & Business Media. doi:10.1007/978-1-4899-0858-2. ISBN 978-1-4899-0860-5.

- Stevens, C. Edward; Hume, Ian D. (1995). Comparative Physiology of the Vertebrate Digestive System. Cambridge University Press. p. 317. ISBN 978-0-521-44418-7.

- Klinowska, Margaret; Cooke, Justin (1991). Dolphins, Porpoises, and Whales of the World: the IUCN Red Data Book (PDF). Columbia University Press, NY: IUCN Publications. ISBN 978-2-88032-936-5.

- Ralls, Katherine; Mesnick, Sarah. "Sexual Dimorphism". Encyclopedia of Marine Mammals (PDF) (2nd ed.). San Diego: Academic Press. pp. 1005–1011. ISBN 978-0-08-091993-5.

- Whitehead, Hal (2003). Sperm Whales: Social Evolution in the Ocean. Chicago: University of Chicago Press. ISBN 0-226-89518-1.

- Thewissen, J. G. M.; Perrin, William R.; Wursig, Bernd (2002). "Hearing". Encyclopedia of Marine Mammals. San Diego: Academic Press. pp. 570–572. ISBN 978-0-12-551340-1.

- Schokkenbroek, Joost (2008). "King Willem I and the Premium System". Emilia%20in%201778&f=false Trying-out: An Anatomy of Dutch Whaling and Sealing in the Nineteenth Century. p. 46. ISBN 978-90-5260-283-7. Retrieved 25 November 2015.

- Stackpole, E. A. (1972). Whales & Destiny: The Rivalry between America, France, and Britain for Control of the Southern Whale Fishery 1785–1825. The University of Massachusetts Press. ISBN 0-87023-104-9.

- Whitehead, H. (2003). "Ghosts of Whaling Past". Sperm Whales Social Evolution in the Ocean. University of Chicago Press. pp. 360–362. ISBN 0-226-89518-1.

- Berta, A.; Sumich, J. L.; Kovacs, K. M. (2015). Marine Mammals: Evolutionary Biology. Academic Press. p. 430. ISBN 978-0123970022.

Extinct Marine Mammals

The sea mink has become an extinct marine mammal; they had a distinctive odor and had fur that was known to be rough. The other extinct marine mammals explained in the chapter are Steller's sea cow, Japanese sea lion and Caribbean monk seal. This section helps the readers in understanding the causes of disappearance of these marine mammals.

Sea Mink

The sea mink (*Neovison macrodon*) is an extinct North American member of the family Mustelidae. It is the only mustelid, and one of only two terrestrial mammal species in the order Carnivora, to become extinct in historic times (the other being the Falkland Islands wolf). The body of the sea mink was significantly longer than that of the closely related American mink (*N. vison*), and also bulkier, leading to a pelt that was almost twice the size of the other species. The longest specimen recorded was said to be 82.6 cm (32.5 in). The sea mink produced a distinctive odor, and had fur that was said to be coarser and redder than the American mink's.

Appearance

The sea mink was hunted to extinction before scientists had an opportunity to analyze them. Its relatives give a general idea of what this semiaquatic weasel looked like. Accounts from locals to the New England/Atlantic Canadian regions say that the sea mink had a fatter body than that of the American mink. Furthermore, it had reddish fur, and its tail was slightly[vague] bushy. This larger body and fur made it very profitable and desirable to fur trappers.

Diet

Similar to the European mink and the American mink, the sea mink's diet consisted of seabirds, most likely the Labrador duck, seabird eggs, hard-bodied marine invertebrates, and in some cases insects. Since the sea mink was larger than the other two species of Mustelidae it is assumed that it ate in greater proportions.

Skeletal Structure

The sea mink was the largest of the minks. Its skull had a wide rostrum, large opening of anterior nares, large antorbital foramina, and very large teeth. This species skull was easily distinguishable from the other species of vison, American mink, because of its large size and bigger teeth. Fur buyers and traders recognized the sea mink because of its larger size compared to other minks and eventually the species was exterminated by the interest in their fur.

Many skeletal remains of this mink have been found off the New England coast. The sea mink served as food for the Native American tribes that once lived there. These minks are described as large and heavily built, with a low sagittal crest and short, wide postorbital processes. Fragmentary skeletal remains of the sea mink leave most of its external measurements to speculation.

Behavior

The sea mink was characterized by its solitary, territorial nature. Males were known to be particularly aggressive towards each other, marking territories with specific scents along a shoreline. In the event of trespassing, violent confrontations would occur. At the same time, males and females both lived promiscuous lifestyles, oftentimes mating with multiple partners especially during late spring, April to May timeframe.

The litter of pups, usually blind and helpless, were supported by the mother for a period of 13–14 weeks. However, many external threats affected the development of the baby pups, and high mortality rates were not uncommon.

Habitat

Sea minks were semiaquatic animals that lived around the rocky coasts of New England and Atlantic Canada, as far north as Nova Scotia in rocky areas along the coastal lines. They inhabited the shores of New England and the Canadian Maritime Provinces until hunted to extinction in the 19th century. They were not a truly marine species, being confined to coastal waters. They also habituated near Casco Bay, Maine, in the south to as far north as Newfoundland, Canada. The Labrador duck, with which it coexisted, may have been a prey item.

The sea mink was the most aquatic member of Musteloidea except for the otters, and it was unusual in having rapidly evolved toward a marine habitat in the late Cenozoic. The sea mink family was originally from coastal Eastern North America, from Massachusetts to the Maritime Provinces. It had a more slender body than the American mink. It preferred coastal habitats, particularly rocky coasts and offshore islands.

Extinction

The sea mink was hunted to extinction to satisfy the demand of the European fur market. Fur traders made traps to catch the sea minks and also pursued them with dogs. Even before the European expansion, Native Americans would capture the animals for their pelts and flesh. A large contributing factor to the eventual extinction of the sea mink was the unregulated hunting and harvesting of these animals. Another possible contributing factor was the high mortality rate of the young. Ultimately, the sea mink became extinct sometime between 1860 and 1870.

Subspecies

Debate has occurred regarding whether the sea mink is its own species, or a subspecies of American mink. Those who argue that the sea mink is a subspecies often refer to it as *Neovison vison macrodon*. However, research has been conducted in which the teeth of the sea mink, dating back 5000 to 250 years ago, were compared with 158 other mink species. It found a substantial difference in dental proportions between *N. macrodon* and *N. vison*, comparable to the dental differences between pairs of related species. The sea minks' back teeth were broader than those of American mink, suggesting the sea mink had adapted over time to eating harder-bodied prey. This suggests reproductive isolation, an independent evolutionary direction, and ultimately, support for the hypothesis that sea minks are their own species *Neovison macrodon*.

Steller's Sea Cow

The Steller's sea cow, *Hydrodamalis gigas*, is an extinct herbivorous marine mammal of the North Pacific Ocean. It was the largest member of the order Sirenia, which includes its closest living relative, the dugong (*Dugong dugon*), and the manatees (*Trichechus* spp.). The Steller's sea cow reached up to 9 metres (30 ft) in length, making it among the largest mammals other than whales to have existed in the holocene epoch. Steller's sea cow was first described by Georg Wilhelm Steller. Although the Steller's sea cow had formerly been abundant throughout the North Pacific, by the mid 1700s, its range had been limited to a single, isolated population surrounding the uninhabited Commander Islands. It was hunted for its meat, skin, and fat by fur traders, and was also hunted by aboriginals of the North Pacific coast. Within 27 years of discovery by Europeans, the slow-moving and easily captured Steller's sea cow was hunted to extinction.

Description and Ecology

The sea cow grew to at least 26 to 30 ft (8 to 9 m) in length as an adult, much larger than the manatee or dugong; however, concerning their weight, Steller's work contains two contradictory estimates: 4.4 and 26.8 short tons (4 and 24.3 metric tons). The true value is estimated to lie between these figures, at around 9 to 11 short tons (8 to 10 metric tons). Their large size was probably to reduce their surface-area-to-volume ratio and conserve heat. The forelimbs, according to Steller, were used as sort of a holdfast to anchor themselves down to prevent being swept away by the strong nearshore waves around their habitat. Unlike other sirenians, the Steller's sea cow was positively buoyant, meaning they could not completely submerge. They had a thick epidermis to prevent injury from abrasions on sharp rocks and ice, and possibly to prevent the skin that was not submerged from drying out.

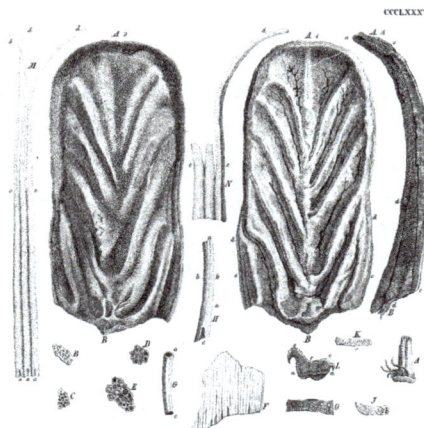

Illustrations of the dentition of Steller's sea cow by Johann Christian Daniel von Schreber

Its head was small and short compared to the huge body. The upper lip was large and broad, and extended so far beyond the mandible, that the mouth appeared to be located underneath the skull. Instead of teeth, Steller's sea cow had a dense array of white bristles, 1.5 inches (3.8 cm) long, which were used to pull out seaweed and hold food, and used keratinous plates for chewing. According to Steller, these plates, or "masticatory bones", were held together by papillae and had many small holes for nerves and arteries. The rostrum was pointed downwards, as in all sirenians, to better grasp kelp. Like other sirenians, the Steller's sea cow was an obligate herbivore, and kelp was most likely their main food source. They may have also fed on seagrasses, but this could not have been a main food source for supporting a viable population, because grasses did not occur in sufficient quantity. Since this animal floated, it most likely fed on canopy kelp. Kelp releases a chemical deterrent to prevent grazing, but canopy kelp release a lower concentration, allowing the sea cows to graze without developing resistance.

Whether or not Steller's sea cows had any predators is unknown. They may have been hunted by killer whales and sharks, but their buoyancy may have made it difficult for killer whales to drown them, and the rocky kelp forests may have protected them from sharks. According to Steller, the young were guarded by the adults from predators.

Taxonomy and Range

The Steller's sea cow was a direct descendant of the Cuesta sea cow, an extinct tropical sea cow of California. It most likely went extinct due to the onset of the Ice Ages and the subsequent cooling of the oceans; lineages which could not adapt died out, and those that could started the lineage of the Steller's sea cow. The Steller's sea cow was a member of the genus *Hydrodamalis*, a group of large sirenians, whose sister taxon was *Dusisiren*. Much like the Steller's sea cow, the ancestors of the *Dusisiren* were associated with tropical mangroves, and adapted to the cold climates of the North Pacific and to consuming kelp.

Steller's sea cow was discovered in the mid-18th century by Georg Wilhelm Steller, and subsequently named after him; Steller studied a relict population near Bering Island while he was shipwrecked there. His account was written in his posthumous publication *De bestiis marinis*, or "The Beasts of the Sea". In 1811, naturalist Johann Karl Wilhelm Illiger placed the Steller's sea cow under the genus *Rhytina*, which many writers at the time adopted. However, the animal had already been classified long before this. Zoologist Eberhard August Wilhelm von Zimmermann had described its specific name as *gigas* in 1780, and biologist Anders Jahan Retzius, 17 years before Illiger, had described the sea cow as *Rhytina* and placed it under the genus *Hydrodamalis*. He, however, described its specific name as *stelleri*, as Steller was the first person to describe it. It was not until the 1900s that *Hydrodamalis gigas* was used.

Their range at the time of their discovery was apparently restricted to the Commander Islands, though fossils dating to the late Pleistocene were found in Monterey Bay, California. The first fossils discovered outside the Commander Islands were interglacial Pleistocene deposits in Amchitka. There is evidence that sea cows also inhabited the Near Islands during historic times.

Extinction

The Steller's sea cow was quickly wiped out by the sailors, seal hunters, and fur traders who followed Vitus Bering's route past the islands to Alaska, who hunted it for its meat before sailing to nearby islands in search of sea otter pelts. It was also hunted for its valuable subcutaneous fat, which was not only used for food (usually as a butter substitute), but also for oil lamps because it did not give off any smoke or odor and could be kept for a long time in warm weather without spoiling. By 1768, 27 years after it had been discovered by Europeans, Steller's sea cow was extinct. It has been argued that the Steller's sea cow's decline may have also been an indirect response to the harvest of sea otters by aboriginal people from the inland areas. With the otters reduced, the population of sea urchins would have increased and reduced availability of kelp, the sea cow's primary source of food. Thus, aboriginal hunting of both species may have contributed to the sea cow's disappearance from continental shorelines. In historic times, though, aboriginal hunting had depleted sea otter populations only in localized areas. The sea cow would have been easy prey for aboriginal hunters, who would likely have exterminated accessible populations with or without simultaneous otter hunting. In any event, the sea cow was limited to coastal areas off islands without a human population by the time Bering arrived, and was already endangered. It is possible that the extinction of these remaining endangered populations of sea cow could have been affected solely by the hunting of the sea cow for meat by fur-trading mariners of the time, and no other factors (such as overpopulation of sea urchins) need contributed. Zoologist Leonhard Hess Stejneger estimated in 1887 that there had been fewer than 1,500 individuals remaining at the time of their discovery, and thus had been in immediate danger of extinction.

Reconstruction

In 1963 the Russian magazine *Priroda* (Nature), official journal of the Academy of Sciences of the USSR, published an article reporting a possible sighting. In 1962 the whaling ship *Buran* had reported a group of large marine mammals in shallow water off Kamchatka, grazing on seaweed. These were considered possibly a surviving sea cow population, but there were no further known reports.

Portrayals in Media

Sea cows give Kotick the necessary information to find a new home in the story *The White Seal* contained in The Jungle Book by Rudyard Kipling.

Tales of a Sea Cow is a 2012 film by Icelandic-French artist Etienne de France "documenting" a fictional 2006 re-discovery by scientists of a population of Steller's sea cows off the coast of Greenland via sound recordings or their calls. This film has been exhibited in public institutions such as art museums and universities in Europe. Art critic Annick Bureaud found the film a "tongue in cheek and joyous but unsettling fable".

Japanese Sea Lion

The Japanese sea lion is an aquatic mammal thought to have become extinct in the 1970s.

Prior to 2003, it was considered to be a subspecies of California sea lion as *Zalophus californianus japonicus*. However, it was subsequently reclassified as a separate species. Some taxonomists still consider it as a subspecies of the California sea lion. It has been argued that *Z. japonicus*, *Z. californianus*, and *Z. wollenbaeki* are distinct species because of their distant habitation areas and behavioral differences.

They inhabited the Sea of Japan, especially around the coastal areas of the Japanese Archipelago and the Korean Peninsula. They generally bred on sandy beaches which were open and flat, but sometimes in rocky areas.

Currently, several stuffed specimens can be found in Japan and in the National Museum of Natural History, Leiden, the Netherlands, bought by Philipp Franz von Siebold. The British Museum possesses a pelt and four skull specimens.

Physical Description

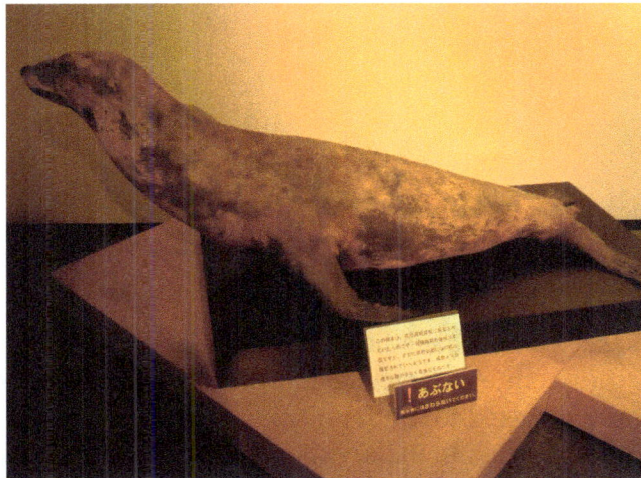

Taxidermied specimen at AQUAS

Male Japanese sea lions were dark grey and weighed about 450 to 560 kg, reaching lengths of 2.3 to 2.5 m; these were larger than male California sea lions. Females were significantly smaller at 1.64 m long with a lighter colour than the males.

Range and Habitat

Japanese sea lions were primarily found in the Sea of Japan along the coastal areas of the Korean Peninsula, the mainlands of the Japanese Archipelago (both sides on the Pacific Ocean and Sea of Japan), the Kuril Islands, and southern tip of the Kamchatka Peninsula.

Old Korean accounts also describe that the sea lion and spotted seal (*Phoca largha*) were found in broad area containing the BoHai Sea, the Yellow Sea, and Sea of Japan. The sea lions and seals left relevant place names all over the coast line of Japan, such as Ashika-iwa (アシカ岩, sea lion rock) and Inubosaki (犬吠崎, dog-barking point) because of the similarity of their howls.

Lifestyle and Reproduction

They usually bred on flat, open, and sandy beaches, but rarely in rocky areas. Their preference was to rest in caves.

Human Uses

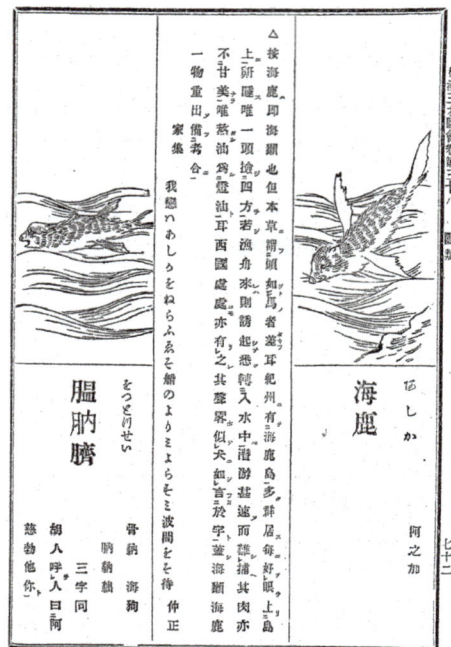

Sea lion (right) and fur seal, Wakan Sansai Zue (around 1712)

Many bones of the Japanese sea lion have been excavated from shell middens from the Jōmon period in Japan while an 18th-century encyclopedia, *Wakan Sansai Zue*, describes that the meat was not tasty and they were only used to render oil for oil lamps. Valuable oil was extracted from the skin, its internal organs were used to make expensive oriental medicine, and its whiskers and leather were used as pipe cleaners and leather goods, respectively. At the turn of the 20th century, they were captured for use in circuses.

Extinction

Harvest records from Japanese commercial fishermen in the early 1900s show that as many as 3,200 sea lions were harvested at the turn of the century, and overhunting caused harvest numbers to fall drastically to 300 sea lions by 1915 and to few dozen sea lions by the 1930s. Japanese commercial harvest of Japanese sea lions ended in the 1940s when the species became virtually extinct. In total, Japanese trawlers harvested as many as 16,500 sea lions, enough to cause their extinction. Submarine warfare during World War II is even believed to have contributed to their habitat destruction. The most recent sightings of *Z. japonicus* are from the 1970s, with the last confirmed record being a juvenile specimen captured in 1974 off the coast of Rebun Island, northern Hokkaido.There were a few unconfirmed sightings in 1983 and 1985.

Sightings of single sea lions of unclear identities have been reported at Iwami, Tottori in July, 2003, and on Koshikijima Islands in March, 2016. Both animals were positively identified as *otariinae* based on photographs.

Population Revival Efforts

In 2007, the South Korean Ministry of Environment has announced that South and North Korea, Russia, and China will collaborate on bringing back the Japanese sea lion in the Sea of Japan. The National Institute of Environmental Research of Korea was commissioned to conduct feasibility research for this project. If the animal cannot be found, the South Korean government plans to relocate California sea lions from the United States. The South Korean Ministry of Environment supports the effort because of the symbolism, national concern, the restoration of the ecological system, and possible ecotourism.

Caribbean Monk Seal

The Caribbean monk seal, West Indian seal or sea wolf (as early explorers referred to it), *Neomonachus tropicalis* (formerly *Monachus tropicalis*), was a species of seal native to the Caribbean and is now believed to be extinct. The Caribbean monk seals' main predators were sharks and humans. Overhunting of the seals for oil, and overfishing of their food sources, are the established reasons for the seals' extinction. The last confirmed sighting of the Caribbean Monk Seal was in 1952 at Serranilla Bank, between Jamaica and Nicaragua. In 2008 the species was officially declared extinct in the United States of America after an exhaustive search for the seals which lasted for about five years. This analysis was conducted by the National Oceanic and Atmospheric Administration and the National Marine Fisheries Service. Caribbean Monk Seals were closely related to the Hawaiian monk seals, which live around the Hawaiian Islands and are now endangered, and Mediterranean monk seals, another endangered species. An estimated 600 Mediterranean monk seals and 1,100 Hawaiian monk seals are alive in the wild.

Description

Drawing of *Monachus tropicalis*

Caribbean monk seals had a relatively large, long, robust body, could grow to nearly 2.4 metres (8 ft) in length and weighed 170 to 270 kilograms (375 to 600 lb). Males were probably slightly larger than females, which is similar to Mediterranean monk seals. Like other monk seals this species had a distinctive head and face. The head was rounded with an extended broad muzzle. The face had relatively large wide-spaced eyes, upward opening nostrils, and fairly big whisker pads with long light-colored and smooth whiskers. When compared to the body, the animal's foreflippers were relatively short with little claws and the hindflippers were slender. Their coloration was brownish and/or grayish, with the underside lighter than the dorsal area. Adults were darker than the more paler and yellowish younger seals. Caribbean monk seals were also known to have algae growing on their pelage, giving them a slightly greenish appearance, which is similar to Hawaiian monk seals.

Behavior and Ecology

Historical records suggest that this species may have "hauled out" at sites (resting areas on land) in large social groups (typically 20-40 animals) of up to 100 individuals throughout its range. The groups may have been organised based on age and life stage differences. Their diet most likely consisted of fish and crustaceans.

Like other true seals, the Caribbean monk seal was sluggish on land. Its lack of fear for humans and an unaggressive and curious nature also contributed to its demise.

Reproduction and Longevity

Two young individuals in New York Aquarium, 1910

Caribbean monk seals had a long pupping season, which is typical for pinnipeds living in subtropical and tropical habitats. In Mexico, breeding season peaked in early December. Like other monk seals, this species had four retractable nipples for suckling their young. Newborn pups were probably about 1 metre (3 ft 3 in) in length and weighed 16 to 18 kilograms (35 to 40 lb) and reportedly had a sleek, black lanugo coat when born. It is believed this animal's average lifespan was approximately twenty years.

Habitat

Caribbean monk seals were found in warm temperate, subtropical and tropical waters of the Caribbean Sea, Gulf of Mexico, and the west Atlantic Ocean. They probably preferred to haul out at sites (low sandy beaches above high tide) on isolated and secluded atolls and islands, but occasionally would visit the mainland coasts and deeper waters offshore. This species may have fed in shallow lagoons and reefs.

Relationship with Humans

Depiction by Henry W. Elliott from 1884

The first historical mention of the Caribbean monk seal is recorded in the account of the second voyage of Christopher Columbus. In August 1494 a ship laid anchor off the mostly barren island of Alta Velo, south of Hispaniola, the party of men went and killed eight seals (Sea Wolves) that were resting on the beach. The second recorded interaction with Caribbean monk seals was Juan Ponce de León's discovery of the Dry Tortugas Islands. On June 21, 1513 Ponce de León discovered the islands, he ordered a foraging party to go ashore, where the men killed fourteen of the docile seals. There are several more records throughout the colonial period of seals being discovered and hunted at Guadelupe, the Alacrane Islands, the Bahamas, the Pedro Cays, and Cuba. As early as 1688 sugar plantations owners sent out hunting parties to kill hundreds of seals every night in order to obtain oil to lubricate the plantations machinery. A 1707 account describes fisherman slaughtering seals by the hundreds for oil to fuel their lamps. By 1850 so many seals had been killed that there were no longer sufficient numbers for them to be commercially hunted.

In the late nineteenth and early twentieth centuries scientific expeditions to the Caribbean encountered the Caribbean monk seal. In December 1886 the first recorded scientific expedition, to research seals, led by H. A. Ward and Professor F. Ferrari Perez as part of the Mexican Geographical and Exploring Survey, ventured to a small collection of reefs and a small cay known as the triangles (20.95° N 92.23° W) in search of Monachus tropicalis. Although the research expedition was in the area for only four days, forty-two specimens were killed and taken away; the two leaders of the expedition shared them. Two specimens from this encounter survive intact at the British Museum of Natural History and the Cambridge Zoological Museum respectively. The expedition also captured a newly born seal pup that died in captivity a week later.

The first Caribbean monk seal to live in captivity for an extended period was a female seal that lived in The New York Aquarium. The seal was captured in 1897 and died in 1903, living in captivity for a total of five and one-half years. In 1909 The New York Aquarium acquired four Caribbean monk seals, three of which were yearlings (between one and two years old), and the other a mature male.

Extinction

Through the first half of the twentieth century, Caribbean monk seal sightings became much more rare. In 1908 a small group of seals was seen at the once bustling Tortugas Islands. Fishermen captured six seals in 1915, which were sent to Pensacola, Florida, and eventually released. A seal was killed near Key West, Florida in March 1922. There were sightings of Caribbean monk seals on the Texas coast in 1926 and 1932. The last seal recorded to be killed by humans was killed on the Pedro Cays in 1939. Two more seals were seen on Drunken Mans Cay, just south of Kingston, Jamaica, in November, 1949. In 1952 the Caribbean monk seal was confirmed sighted for the last time at Serranilla Bank, between Jamaica and Nicaragua.

The final extinction of the Caribbean monk seal was triggered by two main factors. The most visible factor, contributing to the Caribbean monk seals demise, was the nonstop hunting and killing, of the seals, in the eighteenth and nineteenth centuries to obtain the oil held within their blubber. The insatiable demand for seal products in the Caribbean encouraged hunters to slaughter the Caribbean monk seals by the hundreds. The Caribbean monk seals' docile nature and lacking flight instinct in the presence of humans made it very easy for anyone who wanted to kill one to do so. The second factor was the over fishing of the reefs that sustained the Caribbean monk seal population. With no fish or mollusks to feed on, the seals that were not killed by hunters for oil died of starvation or simply did not reproduce as a result of an absence of food. Surprisingly little was done towards attempting to save the Caribbean monk seal; by the time it was placed on the endangered species list in 1967 it was likely already extinct.

Unconfirmed sightings of Caribbean monk seals by local fishermen and divers are relatively common in Haiti and Jamaica, but two recent scientific expeditions failed to find any sign of this animal. It is possible the mammal still exists, but some biologists strongly believe the sightings are of wandering hooded seals, which have been positively identified on archipelagos such as Puerto Rico and the Virgin Islands. On April 22, 2009, The History Channel aired an episode of *Monster Quest*, which hypothesized an unidentified sea creature videotaped in the Intracoastal Waterway of Florida's southeastern coast could possibly be the extinct Caribbean monk seal. No conclusive evidence has yet emerged in support of this contention, however, and opposing hypotheses asserted the creature was simply a misidentified West Indian manatee, common to the area.

References

- Day, David (1981). The Encyclopedia of Vanished Species. London: Universal Books Ltd. p. 220. ISBN 0-947889-30-2.

- Kays, Roland W. & Wilson, Don E. (2009). Mammals of North America (Paperback) (2nd ed.). Princeton University Press. p. 180. ISBN 9780691140926.

- Wilson, Don E.; Reeder, DeeAnn M. (eds.). Mammal Species of the World: A Taxonomic and Geographic Reference. 1 (2 ed.). Baltimore: Johns Hopkins University Press. p. 92. ISBN 978-0-8018-8221-0.

- Steller, Georg W. (2011) [1751]. "The Manatee". In Royster, Paul. De Bestiis Marinis. Lincoln: University of Nebraska. pp. 13–43. ISBN 978-1-295-08525-5.

- Perrin, William F.; Wursig, Bernd; Thewissen, J. G. M. (2008). Encyclopedia of Marine Mammals (2nd ed.). San Diego: Academic Press. p. 306. ISBN 978-0-12-373553-9.

- Domning, D. P. (1978). Sirenian evolution in the North Pacific Ocean. 118. University of California Publications in Geological Sciences. pp. 1–176. ISBN 978-0-520-09581-6.

- MacDonald, Stephen C.; Cook, Joseph A. (2009). Recent Mammals of Alaska. University of Alaska Press. pp. 57–58. ISBN 978-1-60223-047-7.

- Haycox, Stephen W. (2002). Alaska: An American Colony. Seattle: University of Washington Press. pp. 55, 144. ISBN 978-0-295-98249-6.

- Ellis, Richard (2004). No Turning Back: The Life and Death of Animal Species. New York City: Harper Perennial. p. 134. ISBN 978-0-06-055804-8.

- Domning, D.; Anderson, P.K.; Turvey, S. (2008). "Hydrodamalis gigas". IUCN Red List of Threatened Species. Version 2008. International Union for Conservation of Nature. Retrieved 19 August 2016.

- Lowry, L. 2015. Neomonachus tropicalis. The IUCN Red List of Threatened Species 2015: e.T13655A45228171. doi:10.2305/IUCN.UK.2015-2.RLTS.T13655A45228171.en. Downloaded on 03 December 2015.

- Canadian Wildlife Service. (2006). Sea Mink. Environment Canada – Species at Risk. <http://www.speciesatrisk.gc.ca>. Downloaded on 19 October 2014.

Permissions

Index